# Gold In Trib 1

Flying, Hiking, and Gold Prospecting Adventure in Wild, Present-day Alaska.

*Douglas Anderson*

## Douglas Anderson

PO Box 221974 Anchorage, Alaska 99522-1974

ISBN 1-888125-11-X

Library of Congress Catalog Card Number: 96-71689

Copyright 1997 by Douglas Anderson
—First Edition—

All rights reserved, including the right of reproduction in any form, or by any mechanical or electronic means including photocopying or recording, or by any information storage or retrieval system, in whole or in part in any form, and in any case not without the written permission of the author and publisher.

Manufactured in the United States of America.

Gold in Trib 1 is dedicated to the miners, prospectors, and explorers who hunted Alaska for gold before maps, planes, trucks, or polypropylene. These brave men have my highest admiration.

# Table Of Contents

| | | |
|---|---|---|
| Foreword | | 7 |
| Ladue Area Map | | 9 |
| Introduction | | 11 |
| 1 | The Plan | 13 |
| 2 | Hatcher Pass | 23 |
| 3 | Practice Air Drop | 29 |
| 4 | Powerline Trail | 35 |
| 5 | Polka Night | 45 |
| 6 | Flight to Gulkana | 49 |
| 7 | Air Drop | 55 |
| 8 | Return to Anchorage | 63 |
| 9 | The Drive | 69 |
| 10 | Up to the Ridge | 79 |
| 11 | Day One on the Trail | 85 |
| 12 | Wet Saturday | 95 |
| 13 | Trib 1 at Last | 101 |
| 14 | Trib 2 | 111 |
| 15 | Glory Hole | 117 |
| 16 | A Day of Panning | 121 |
| 17 | One Mean Wet Hike | 133 |
| 18 | Last Day of Hiking | 141 |
| 19 | Hagen's Winter Trip | 147 |
| 20 | Working Gold Claim | 167 |
| 21 | Day of Relaxation | 177 |
| 22 | Last Few Days | 183 |
| 23 | Back to Civilization | 187 |

# Beginning

## Foreword

> I wanted the gold and I got it,
> Came out with a fortune last fall,
> Yet somehow life's not what I thought it,
> And somehow the gold isn't all.
> *Spell of the Yukon—Robert Service*

**R**eading Gold on Trib 1 reminded me of those adventuresome years spent with my good friend, Doug, searching for gold. It seemed easier then to take risks and choose between a "do and dare" adventure on one hand, and "conformity and a secure pay check" on the other.

Everyone hopes there will be a pot of gold at rainbow's end, and likewise, great wealth at the end of a prospecting adventure—a Glory Hole. For some, it turns out that way.

Other than color in the pan, or a nugget or two in the poke, what is there for reflection? Our efforts were great and investment substantial, but we do not balance pleasure against despair, gains against losses. Instead, I look back with a sense of pride on a journey taken, hardships endured, and an occasional success over the devil in my soul.

What is life if not a journey, a taking of risks, and occasional successes? For the faint hearted, a journey not taken is but a life not lived.

Gold in Trib 1, is a fitting tribute to our journey, a reminder of our lives not frittered away, even though sometimes it seemed to be so. A host of memories are a lasting reward more valuable than the gold we discovered. *Hagen Gauss*

Doug (on left) and Hagen on the trail at Squirrel Peak.

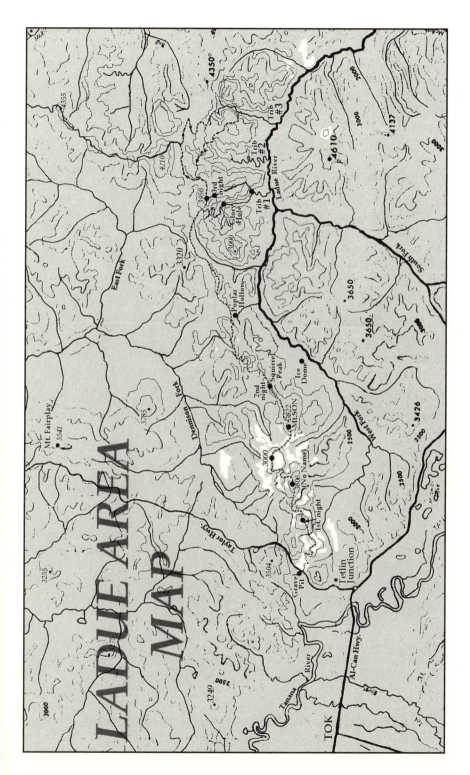

# Preface
## Introduction

**H**aving made Alaska my home for ten years, I often thought I had enough material to write a book. Initially, I hand wrote sections of the manuscript, over a two year period, while traveling as a company representative. Long airplane flights and many lonely nights in motels in strange places were eased because I became engrossed in my manuscript. Indeed, there were times when I was reliving those events which really happened: I was flying in my plane, hiking those high mountain ridges, or panning for gold. It seemed at times that I could not write fast enough, there was just so much material to work with, so many ideas spilling forth. I thought that writing would be great therapy and would help to relieve my desperate feelings of home sickness for Alaska. It did not—though writing served to pass the time—I was even more homesick, but I derived much pleasure from penning the tale.

Gold in Trib 1 is an account of the exploits of myself and a very good friend, and, in particular, of our adventures while prospecting for gold. It is a factual account where possible and where not factual, is the way we would have liked it.

To make the book more appealing to readers I had, of

necessity, cause to adjust some chronology and to take license with regard to some details which might otherwise have been boring. As a result, it is my fervent hope that readers enjoy the book for what it is, and will not take it so seriously as to dash off to the subject area with expectations of finding their fortune, though I know from experience they would find it a very challenging wilderness hike. Naturally there is still much gold in Alaska, but I fear I may have made discovering the Glory Hole, wherever it may be, sound somewhat easier and more financially rewarding than it really was.

## Chapter 1
## The Plan

My red and white Cessna 150, throttled back to a soft burbling seventeen hundred rpms, and with carb heat selected, was in a gentle left turn. Wasilla's landing strip was in sight under my left wing tip.

"No reported traffic," the helpful attendant at the Palmer Airport Regional Flight Service Station (FSS) reported through the 150's radio receiver. I knew he didn't mean there wasn't any traffic. It just meant he wasn't aware of any from his location, ten miles away in Palmer. Wasilla's airport is not strictly controlled and pilots are not obligated to use a radio at all.

In my six years of flying in Alaska, I had always tried to fly my little plane as professionally as possible, though I held only a private pilot license. The Cessna 150 was equipped with a radio and a VOR as aids to navigation and I used them as much as practical, given the circumstances.

Customarily, I contacted the FSS at Palmer to state my intention of landing at Wasilla and the attendant responded with a report of the situation as he knew it. From then on, I relied on my own eyes to make sure there really wasn't any traffic in the area.

I flew over the airport at fifteen hundred feet and looked

for air traffic. Seeing none, I circled, slipped into the downwind leg, parallel to the runway, dropped one notch of flaps, and turned gently onto base leg at five hundred feet. Leveling the wings, I took another good look around. There were no aircraft in sight except those parked by the service station and cafe at the west end of the airport. Wasilla hasn't any fancy aids of larger airports: approach indicator lights and the like, so all approaches and landings are simply by pilot judgment. I called Palmer FSS again and advised them I was going ahead with my approach and landing to the west.

The wind sock at the service station suggested a light wind from the southwest, so I would have to cope with a quartering crosswind during the landing. Nudging the electric flap control lever down one more notch, I turned the plane onto final approach in line with the runway. I dropped the flaps further, slowed the aircraft to about fifty eight miles per hour, and trimmed the engine speed and controls to maintain a gradual descent. The plane crabbed sideways due to the crosswind, but was still on a course exactly in line with the runway.

Tops of silver birch trees, tipped with fresh green leaves, slid by under the 150's wings. A gravel road and chain link boundary fence came into view, and then the threshold of the runway. Applying a steady back pressure to the yoke, I eased the throttle back and the aircraft touched down with a cautioning squawk from the stall warning indicator. There was a noisy rumble as the main gear contacted the runway and loose gravel rattled off the underside of the fuselage. I maintained steady back pressure on the control yoke to keep the nose wheel out of ruts and to protect the propeller from gravel rash. With my right hand, I flipped up the electric flap switch and switched off carb heat and the landing lights.

The last thirty seconds had been busy as I made the final approach and landing. Now I had time to glance to the right toward my friend Hagen's house, my destination for the weekend.

Hagen's house, part single level, part two level, stood on

a gentle rise north of the airport boundary. It was a pilot's dream house. A plane could taxi from the runway, cross the gravel road, and park on a level area by the house. Last year we buried three attachment anchors for aircraft tiedown ropes. Hagen's was not the only residence on this side of the runway, and several neighbors had aircraft on their property. One recently completed a large hanger.

Not seeing aircraft on final approach, I turned the 150 around and taxied back along the runway to a driveway leading to Hagen's house. Looking both ways for traffic, I taxied across the road and up Hagen's driveway to the grass covered tiedown area. A tight one-eighty turn and the plane was in line with the anchor points.

Hagen watched my approach and was standing on the wooden deck spanning the full width of the house. In stature he and I were the same, five feet ten inches and one hundred ninety five pounds. Clothing size also the same. The similarity ended there. I tend to be a bit rounded and usually struggled against extra pounds. Hagen was all muscle and sinew with hardly an ounce of fat. Under a shock of curly black hair, with no sign of gray or receding, were lean, angular features. His piercing, blue eyes were deep set under dark eyebrows. Sometimes, he sported a neatly trimmed mustache as he did now. More often than not, Hagen looked serious—in keeping with his Germanic upbringing perhaps—and this invariably gave people the impression he was unfriendly. The opposite was true; Hagen had a great sense of humor. When he smiled, his whole countenance lit up. This was one of his attributes so, when situations warranted, I'd nudge him discreetly and mutter, "Smile damn it."

Hagen raised his hand in greeting, and smiled, as I killed the engine and switched off the navigation lights, radio, and main electrical power. I opened the door and he walked over, greeting me with a cheery "Morning Doug." Still smiling, he took my small nylon overnight bag from me. I installed the control yoke lock and climbed out, locked the door, and secured the plane at three points with the tiedown ropes. It was less than thirty minutes since I had taken off from runway 32 at Anchorage International Airport.

The previous summer I made this trip a dozen times—usually Friday after work. Hagen was carrying out extensive remodeling of his house and I helped whenever possible. The house started as a small one bedroom bungalow but we jacked up the existing structure, poured a completely new foundation, and built a two-story extension, partly overlapping the existing structure. The house was now two thousand square feet of living space and was really too large for Hagen who, like me, was a survivor of divorce and enjoying his single life. I doubted Hagen would ever part with this property because he liked the location so much. Besides, he had put in many hours to make the house just the way he wanted.

During the construction months we established a routine of going to The Pantry Restaurant for many of our meals. We agreed to go there this morning. Hagen deposited my bag in the house and we climbed into his Jeep Cherokee parked in the open garage. As he pulled into the driveway, he pressed the button of the transmitter to close the garage door and we headed toward town.

Wasilla evolved, rather than developed, into a queue of general stores, restaurants, and gas stations strung haphazardly along a half mile stretch of the Parks Highway. The area experienced a boom in the late seventies and early eighties and was tidied up considerably in the process. The main highway through town now had four lanes with turn lanes in the center. Most parking areas in front of the buildings were paved. The town was still far short of beautiful but, was very much improved. Hagen drove to the paved parking space by The Pantry and we walked, between parked vehicles, the short distance to the entrance.

In many ways Hagen was a man of habit. Today was no exception. As I could have predicted, he pumped a couple of quarters into the Anchorage Times paper box and pulled out the thick Saturday morning edition. He always studied and fiercely debated front page politics. He never caught on that I, and others, often wound him up deliberately by an opposing opinion. Nonchalance about an issue was also enough to get him going. Before we were seated, a point in the headlines grabbed his attention and I was already fanning the flames by taking an opposing stance.

Our favorite waitress—cute enough to make us both forget politics—greeted us cheerily with an exaggerated British accent as she tried to mimic my genuine one. She poured our coffee with a flourish without even asking if we wanted any.

In reply to her question, "The usual, old chap?" We said, "yes."

Hagen broke from his paper long enough to give her the eye and to razz her a bit, then was back into page one with the comment, "Listen to what those jerks in Washington are proposing now." I set a trend of tactfully avoiding further comments. When our meal arrived, Hagen set aside the paper, waded in with his usual gusto and was back into section one in no time. He always had a voracious appetite for food. If he hadn't been so energetic, I think he would have had a serious weight problem. As it was, he stayed disgustingly lean and fit. I, who always fretted about the kind of food I consumed, and the calories it contained, struggled with my weight. I ate more slowly and sometimes used it as an excuse not to respond at all to some of Hagen's comments.

We had no firm commitments for the weekend so we took our time over extra coffee and engaged in a lengthy conversation with other regular patrons. At 10 o'clock we paid the tab and left the restaurant. Hagen, having finished with the "meaningful" parts, left the entire paper folded neatly on the table. Someone would enjoy it now if they wished. If I were careful, he might not mention politics again until the next newspaper got him cranked up.

The sunshine was warm and bright. Breakup arrived early in Southcentral Alaska this year. It was now the end of May and the surface snow was melted away and ice was gone from lakes hereabout. There was however, still deep snow on the higher levels of the Chugach Mountains, clearly visible from Wasilla. Typical of Alaska, shady spots would be chilly for another month. Much of the warmth we felt came from being in direct sunlight.

We decided to take a short drive to the Little Susitna River which passed about six miles north of Wasilla. There was a picturesque spot where the road bridge crossed the river. It was a quiet place to enjoy the sunshine. We also wanted to discuss our plans for the summer.

Hagen drove the Jeep northward along the gravel road at a steady fifty miles an hour. Loose gravel popped and pinged from under the wide tires and rattled off the underside of the vehicle. Our speed seemed to iron out most of the washboard effect of the road. The surrounding flat land supported a variety of trees and they were beginning to look green. The leaves of the silver birch, aspens, and alders in particular seemed to burst forth in the space of a few days. Clearings were covered with dwarf willows and other lower growth shrubs. Moose browsed on them during the winter. It was common to see moose in yards, on roads, and in meadows especially when the snow was deep on higher elevations. They migrate to higher levels as snow melts away.

We arrived at the bridge and, as we expected, the river was swollen with melting snow from the mountains. The water was sparkling clear and looked cold. Hagen drove over the one-lane-wide trestle bridge and turned left onto a wide gravel bar. We were the only visitors in this popular summer picnic spot. We left the Jeep and walked back onto the bridge to look down into the stream. There were no fish, except maybe grayling or trout lurking in the eddies downstream of the larger rocks. Soon there would be several runs of spawning salmon. From right here, we'll be able to count hundreds of fish.

We strolled for a short distance downstream along the south bank. Then we clambered over washed up trees, sand bars, and banks of gravel, which had changed shape and location since last summer. During Spring breakup, ice dams and force of water rearranged many things in its path. Many serpentine rivers and streams cut new paths during breakup. They left large, crescent shaped sloughs and ponds where the previous year's river flowed.

In a sandy patch, there was a large fallen tree trunk. It was firmly anchored with its roots buried under a mass of boulders and gravel and its remaining upper branches swept down stream. It was a pleasant spot to lounge, enjoy the sunshine, and talk.

Hagen and I were really different in personality and character. He was from Germany and I'm from England and it gave us something in common. We were both immigrants who became U.S. citizens. We struggled to make ends meet in our new homeland and it was our stubborn European work ethic which pulled us through tough times. We experienced trials and tribulations of divorce, but now found ourselves in the enviable, if not well deserved, position of being "unencumbered." We felt in control of our lives and able to come and go as we pleased.

We shared many common interests and in our free time, joined on many ventures, skiing, hiking, fishing, and on projects like his house. One summertime interest we pursued with tremendous zeal was prospecting for gold.

This was, after all, Alaska. Gold played, and still plays, a significant role in the economy of the State. Gold was avidly searched for by many, some more successfully than others of course. The rewards for much effort could be zero. If lucky, you could make a good living searching for gold.

Some years ago Hagen "caught the bug" and carried out preliminary research on several likely places. One of these was a very remote area north of the Al-Can Highway near the Canadian border. The "gold bug" eventually caught me. As a result, last summer we flew over three valleys and took a series of photographs. During winter months we pieced together the enlarged photo's and now had a photo mosaic.

We were proud of our aerial survey and had a good idea of what the prospect area was like. We thought we understood the challenges of going into this remote place, about forty miles from the nearest road.

We were resigned that we'd have to hike in for the initial look around and we were planning to do just that in eight weeks. We had hiked a dozen times before and knew the importance of careful advance planning. We agreed that, to reassure ourselves about this challenging trip, we needed to fly over the proposed hiking route one more time.

Hagen suggested dropping tools and supplies from the air during this flight and I thought it an excellent idea. Our backpacks would be lighter and without the weight, our traveling would be easier.

Backpacking heavy items like shovels and pick axes for such a distance was out of the question, yet we needed them to prospect effectively. We had no experience dropping anything from an aircraft, so we decided to practice the next day by dropping a package nearby.

We both were forty-three, but physically fit in our own way. We led active lives: hiked, canoed, and fished in summer, skied cross country and downhill in winter. When we were not engaged in outdoor activities, we participated in aerobics, country western and polka dancing. Sometimes, it seemed, there was simply not enough free time from our regular work to do the things we wanted.

Hiking to where we thought the "gold bug" was, would be an ambitious hike. We'd have to walk forty miles, with backpacks, to reach our prospecting site. There was an old trail we would follow along the high ridges for part of the way. We figured three days to walk in, six or seven days to explore and prospect, and two days to walk out. Our earlier hikes paled by comparison.

Exhausting this subject of advance planning, we lounged, facing the sun, our backs against the tree for about an hour. The warmth of the sun soaked us through. Hagen suggested we look at a possible site where we could practice air drops.

We rose, stretched ourselves and walked back to the Jeep. It was unbearably warm inside with the sun blazing in through the windows. We cranked both side windows down as we pulled off the gravel bar and onto the road. Things soon cooled down.

In ten minutes, by way of the back roads, we reached the place Hagen had in mind for the practice drop. It was accessed by a little-used gravel road and was representative of our planned prospect site, except there was no nearby stream. We walked across a hundred yards or so of coarse grass to a slight rise in the terrain. This looked like a good place to practice. There were only a few medium sized spruce trees scattered over the area and plenty of open ground. After looking around, we decided to make our target a distinctive clump of spruce trees. If we flew parallel to the road, we'd easily spot them because they stood in an otherwise clear area. It was not necessary to hit the trees with our package. We only needed to come close to prove our "bombing" technique.

Like many aircraft in Alaska, my 150 was modified to make removal of the right door easy. It would only take a few minutes to prepare the plane for the practice air drop the next day.

## Chapter 2
## Hatcher Pass

F ish Hook Road leads to Hatcher Pass. Being only a few miles from there, we decided to drive into Hatcher Pass and look around. We hadn't been to the pass since March when cross country skiing with friends.

Hatcher Pass is a spectacularly scenic place any time of year. It is a good place to take visitors for a taste of Alaska wilderness. It is possible, during the summer months, to drive all the way over the pass emerging near Willow, a small town on the Parks Highway. There was a good chance the road was still closed at higher elevations with avalanche snow or erosion from melting snow.

We wouldn't attempt to drive too far today, only to the cross country ski area near an Alpine lodge and restaurant. It was late spring but there would still be a few visitors trying to ski the last bit of available snow.

The road approached the pass by crossing a bridge over the Little Susitna River and then turned to follow the course of the river. The road was gravel from the bridge onward but maintained up to the ski area. Almost immediately past the bridge there was a natural gateway formed by a rocky cliff on the left and a towering column of rock on the right. Past this landmark, we were within the boundaries of Hatcher Pass.

In the pass, the river was very different to the quiet and smooth flowing stream where we rested. Here, water was angry, foaming white as it dashed pell-mell through confined channels and over large, well-rounded rocks and against massive boulders. The trees grew at the water's edge. There were a few places to park a vehicle and follow a short pathway to the river's edge. Five miles up the valley road, the tree line stopped and willows and alders took over. Above four thousand feet, terrain was typical arctic tundra with only low growth shrubs and lichens clinging tenaciously to the surface.

The best times to visit the pass was fall when there was a profusion of color. We had both taken some beautiful photographs of the river and valley in years past and had framed several enlargements of which we were quite proud.

We saw many active placer mines, at the lower levels, shut down for the winter months but were just now being reactivated. The whole area is spoken for and it is now impossible to stake a new claim. Most mines are sited on the ancient benches where the river course used to be. Rumor has it some mines are quite productive. Miners jealously guard their claims. It pays to be careful when hiking this area. There were reports of frontier style justice being used on occasion. A few areas were clearly posted and set aside for visitors to try their hand at panning. We had tried here but found only gold dust.

The road began to show signs of deterioration further up the pass, and there were places where runoff from melting snow caused deep washes. The Jeep negotiated these easily but Hagen moaned about getting his clean vehicle dirty. We pressed on, climbing above the tree line, until we reached a junction where the left branch headed on up the pass and the right led to the ski area. High on the mountain side ahead we could see old buildings and remains of historic Independence Gold Mine.

In years past, Independence Gold Mine had been a sizable hard rock operation. The tailings from shafts driven deep into the mountain were clearly visible. They looked like large terraces fanning out below ruined buildings. One large building, originally a dormitory for mine workers and recently renovated, was open in the summer as a museum. We took the road to the right, and, in a couple of minutes reached the parking lot by the ski lodge.

On a good weekend, in the prime skiing season, this parking space was packed with vehicles. Today, only five vehicles nosed in to melting snow banks. Snow cover was visible in the pass, but conditions were not very good for skiing. Anyone out today was likely to be enjoying sunshine and spring warmth more than anything.

Hagen parked the Jeep in a spot clear of breakup mud and we picked our way carefully along the pathway to the ski lodge and restaurant. The building wasn't fancy but sturdily built from rough sawn lumber and trimmed logs. The high A-frame end, facing the valley, was glass and afforded an excellent view. As we entered the restaurant we saw only four people seated in the lounge area by the windows. They were dressed for cross country skiing, but, when asked, confirmed conditions were very poor.

I wasn't very hungry but Hagen was "starving" so we ordered hot chocolate, a cheeseburger for me and hamburger and fries for Hagen. Then, we relaxed by the window until our snack arrived.

As we ate, we contemplated the scenery. "The Chugach mountains are twenty miles away and we are planning to walk forty miles." I said. "It puts scale to what we're about to attempt."

"Seems rather sobering," Hagen replied between bites of his hamburger. "So, the distance might be the same, but the terrain will be different, very different. We just need to be careful. We don't want any disasters."

Our conversation digressed to other subjects and almost an hour passed before we paid the tab and left the restaurant. We climbed into the Jeep, mindful not to carry too much mud into the interior and pulled out of the parking lot. Hagen drove cautiously down the slippery, muddy surface. Hairpin curves regularly claimed a few vehicles. There were no guard rails, and at several places the edge was defined only by a forty-five-degree slope to the valley floor. Going off the road at one of these places was guaranteed to quickly ruin a day.

Hagen was not a cautious skier. These were not recognized downhill ski slopes. He, however, blasted down them last winter, emerged at the bottom plastered with snow. Evidence of close encounters with bushes and dwarf willows still clung to his skis. Several times I left him at the top of the slope and drove to the bottom to meet him. Each time I fully expected him to be a medivac case but he survived unscathed.

We took Fish Hook Road directly to Wasilla stopping first at the self service car wash. Feeding quarters into the hungry coin box, we gave the Jeep a good, high pressure, soapy wash and a rinse that removed all traces of Hatcher Pass mud. We dried the vehicle using a couple of old towels Hagen kept in the back. Hagen drove slowly back to the house, so as not to raise any dust.

Starting the electric door opener, he drove up the driveway into the garage. Unlike many garages, which was a resting place for everything people wanted to keep but never get around to using, Hagen's garage was, clean, tidy, and everything had its place. It was considerably larger than a standard two car garage and there was a large workshop extension at the rear. Hagen built large cabinets against one wall, providing ample storage space for ski equipment, hiking and camping gear, and fishing gear. There was also, a neat place for miscellany not requiring warm storage. The garage was not heated. Installed in the corner by the doorway was a heavy duty fan heater that took the chill off when necessary.

From the cabinets, Hagen selected a variety of tools, which I carried over to the work bench. Soon we had a shovel, pickax, a long handled ax, and a thirty-inch bow saw. We checked the dimensions of the tools against the interior of the Cessna. They'd fit vertically in the passenger compartment and fit into the baggage space behind the seats.

Delving into the cabinets again Hagen came out with a bundle of coarse burlap. We wrapped the tools tightly in a bundle secured with two-inch wide duct tape. Hagen slipped off his boots and popped into the house and reappeared a moment later with a set of bathroom scales. The bundle weighed less than forty pounds. If we could drop these on the prospect area, we'd have tools with which to work.

We speedily removed the right door of the plane. Hagen slid the passenger seat back as far as it would go and climbed in. I lifted the bundle and wedged it inside the door frame along side Hagen's legs. Hagen hugged the bundle close to keep it clear of the control yoke then proved how he could manhandle the bundle and pitch it outward and downward without interfering with any part of the aircraft. It seemed a practical idea. Our trial flight and airdrop the next day would tell us for sure. The door was easily managed. We replaced it, secured the aircraft, tidied up the work bench and then went into the house.

Hagen said he wanted to get married again some day. The problem was he, and I too for that matter, had adapted to single life. We seemed to be too busy pursuing other interests to do any serious lady chasing.

Hagen was quite proud of his house and kept it very tidy. First, the connection between the garage and the house was a sizable mud room. Boots and shoes were placed on built-in shelves while sitting on a bench under a sunny window. Next came the kitchen, followed by the living room and dining area in a traditional L shape. To the back of the house was another large room, designed to be a den, but Hagen used it more as an office. Everything was all very clean and tidy, shattering the common image of frontier lifestyle, and, in particular, the Alaska bachelor lifestyle.

Hagen liked a touch of class, too. The living room and dining room walls supported a fine collection of expensively framed paintings, most of which depicted scenes of Alaska. He particularly liked pioneer scenes and American Indians. There were also a couple of Remington bronze statues and some Alaska Native soapstone carvings on display. Upstairs in the master bedroom were five framed commemorative prints of the Iditarod dog sled race, held each year. Many of Hagen's paintings had appreciated in value since purchase and had been a good investment and a source of pleasure. What he had was tasteful and he was justifiably proud of his collection and his home.

Another surprise, to most people, was that Hagen was an exceptionally good cook. He had a knack for putting the tastiest meals together with whatever was on hand.

We lounged around for an hour watching a little TV, sipping 7-UP, and chatting about a variety of things before Hagen decided he was getting hungry and started to prepare dinner. I knew better than to help so I continued watching TV until Hagen called me to the table.

Dinner turned out to be a tossed green salad and spaghetti bolognaise loaded with meat. It was one of Hagen's specialties with just the right amount of spice. There was crusty French bread warmed with a hint of garlic butter and for dessert some fresh fruit salad. A glass of wine added "that touch of class" we both appreciated.

When we finished eating, I helped clear dishes from the table. Hagen would accept no further help. He had his own routine and was quite intolerant of anyone else in his kitchen. One more reason why finding a suitable companion might prove difficult in his case.

After the cleanup was finished, we watched a good Clint Eastwood movie until eleven and then decided to turn-in. Hagen retired to the upstairs suite while I showered in the smaller downstairs bathroom. I made up a bed on the couch in the living room.

In spite of the size of the house, Hagen furnished only the master suite of upstairs rooms. Before retiring, I had a few chapters of a paperback novel to read so I settled down with my book. I drifted off to sleep with the book on my chest, rolled over later just long enough to switch out the light.

## Chapter 4
### Powerline Trail

Sunday morning dawned with the raucous sound of a plane taking off and we were forced to enjoy an earlier part of the morning than we had planned. It was however, a beautiful morning with a clear view of the snow capped Chugach Mountains from the living room window. We were ready in no time at all and drove to the Pantry for a leisurely breakfast. By chance, a good friend, Joe, was there having breakfast so we joined him. With Joe around there was additional camaraderie. He never let up on razzing the waitresses as they returned. One day, in retaliation, Joe ordered a burger and, when he bit into it, found it dressed with rubber bands for onions. Joe was one of those people who, knowledgeable in politics, could really bring Hagen to a boiling point. Joe brought a measure of character to the restaurant, and, as he did this morning, made it a lively place.

We finally pried ourselves out of there and by eleven-thirty had the plane prepped. We took the door off and Hagen was wedged into place along with the tool bundle. I climbed in, checked the controls, and started the engine. After a few minutes of warm-up, we taxied down the driveway and held just short of the runway. I did an engine run-up check and contacted Palmer FSS. I collected the latest status including barometric pressure, ambient temperature, and wind. It was a balmy sixty one degrees. Hagen spotted a Super Cub on downwind so we waited a few minutes. It took a complete landing and taxied out of the way and then we taxied to the eastern end of the runway.

With a last look around, I applied full power and commenced the take off roll. The Continental engine's crackle assaulted our ears. Gravel pinged from under the tires and had Hagen cringing in his seat. As we lifted off the gravel surface, the noise level reduced dramatically. The engine note was high until I leveled the plane at one thousand feet and reduced the rpm to twenty one hundred.

Mindful of Hagen nervously seated by the open door, I commenced a gentle right turn and started to follow the country road north toward the selected drop site. It took a little over five minutes to get there. Amazing!

Our choice of target was a good one. A half mile to the nearest house and the clump of trees we had selected was easy to locate. Immediately, I coaxed the plane into a descending left turn and lined up with the target. I concentrated on my flying. "Hagen," I shouted. "It's only a trial run so don't do anything foolish." I tried to visualize the trees as the threshold of a runway and flew as if for a landing. Dropping the flaps to thirty degrees and selecting carb heat, I slowed to about fifty eight miles per hour, and trimmed for a gentle glide slope. I adjusted the engine revs and controls to keep us on track and was rewarded by the sight of the trees flashing by twenty feet below the wheels. Quickly I switched off carb heat, piled on the power, and climbed away to a safer altitude.

Hagen yelled, "Let's get this over with!" and clutched his burlap bundle closer to his chest.

I made another gentle left 360 and lined up again with the target. This time Hagen sat up a little and eased the bundle out into the slipstream. Now he could see the approaching target. He'd decide the best time to heave the bundle outward and downward away from the plane. Carefully I tried to replicate the last approach. I sensed, rather than saw, Hagen wrestling the bundle. The aircraft lurched upward. There were no nasty bumps so I knew the drop had been clean, and, most important, Hagen was still firmly in his seat. By this time, I was busy making sure we didn't end in the trees. Only when we were back to a safe altitude did I relax.

I didn't realize how stressful this was. My back was wet with sweat. Low flying, where there was no runway, was for the birds.

Now it was time to head back to Wasilla, pick up our tools and check their condition. I didn't consciously look around for other aircraft during our "bombing" run. It kind of scared me. Now there were several small planes in sight.

Wasilla had a couple of aircraft in the pattern so I raised Palmer on the radio, advised my intention to make a full stop landing, and then joined in the procession on the downwind leg. The plane ahead was practicing touch-and-goes, and didn't stop on the runway. It was lifting off again when I turned onto final. I had plenty of time to land and taxi back to Hagen's driveway before it completed another circuit.

We quickly tied down the plane and refit and locked the passenger side door. Planes are not the most secure vehicles at the best of times but it was the best we could do.

Hagen was quite ecstatic about the air drop. We quickly piled into the Jeep and took off. We zigzagged through the back roads, took the shortest route, and made it in about fifteen minutes. Parking in the same spot as the day before, we headed toward the clump of trees.

"I have a good idea where the bundle landed." Hagen said. "It is to the north side of that patch of open ground."

Sure enough the bundle, still in one piece, was up against a clump of willows. We pulled it into the open to examine it. It survived the impact. The only damage was a large hole cut in the burlap by the corner of the shovel. Tracking back, we found the point of impact, a mini crater in the soft ground. There was a second one five yards closer to the bushes. It looked as if the bundle traveled horizontally as it hit the ground.

Hagen brushing dirt and grass from his pants said, "It would be better to make the drop from a higher altitude. Safer for us too." We'd do that in the real air drop.

I hefted the bundle onto my shoulder, carried it to the Jeep and plunked it into the cargo space. Hagen had the space boxed with tri-wall cardboard to protect the body panels from just such freight. We decided to inspect the tools closely at the house.

In the garage, we unwrapped the tools and examined them carefully. The only damage we could find was that the tube of the bow-saw was bowed more than originally intended and the blade was no longer under tension. This would be easy to fix, even if it happened on the next drop. Our practice air drop had been a success.

Hagen suggested, "We could tie some long streamers of surveyors fluorescent colored tape to the top of the bundle so that we could find it easily."

"Good idea," I agreed. It is surprising how different everything looks at ground level compared with flying. Later, we'd fly to the Ladue Valley and drop the tool bundle and some packages of canned goods. We'd select a place where they could easily be recovered. The site would probably be along our hiking route.

It was one o'clock already so we headed for the Country Kitchen Restaurant where there was always a very good Sunday brunch. Word had spread about the good food and the parking lot was full. We waited only a few minutes for the crowd to thin and got a table by the window with an enjoyable view of the snow capped Chugach Mountains.

It was a buffet brunch and we made several trips to load up our plates. Hunger wouldn't attack us again for a long time. We relaxed, took our time. The restaurant became quieter as we dawdled. Only a few diners remained.

Our talk, although it drifted occasionally to other topics, centered on plans of hiking into our gold prospecting site.

We rarely tired of the subject. I suppose, in the back of our minds, there was always a vision of gold nuggets, big as a fist, gleaming in the bottom of our pans.

Second weekend in July was decided for our air drop and to scout the trail. Of course, as far as flying was concerned, we considered the weather and were prepared to change our plans accordingly. We also decided to take a 'shape up' hike the very next weekend. We'd do this no matter what the weather.

Hagen said, "The chance of fine weather during the entire hike is too much to hope for. We've got to manage whatever is handed out once we're on the way." He was right.

After our long brunch, we lounged around the house. At five thirty I said goodbye and taxied the 150 to the service station to top up the fuel tanks. There was no traffic in sight and Palmer wasn't reporting any so I was able to take off without any delay. There was no wind so I took off to the east. Just as I lifted off, I glanced left to see Hagen on the front deck with one hand raised in a goodbye salute just as he had done each time I left his house.

Twenty-five minutes later the 150 squeaked down on the black tire rubber marks at the end of runway 14 of Anchorage International. It took five minutes to taxi to my rented tiedown spot, close by the control tower. In a few more minutes I had the plane tied down and the flight recorded in my log. My Chevy Blazer was parked nearby and I was soon on the road home.

Home was fifteen minutes away. As I drove to lower hillside area, southeast of town, I mulled over the weekend—all part and parcel of the preparations leading to our big hiking and prospecting adventure.

## Chapter 4

# Powerline Trail

The City of Anchorage, Alaska's largest population center with about two hundred thirty thousand people, is in a valley of ancient glacial moraine. The city, constrained in a roughly triangular area, is bounded by the waters of Knik Arm to the northwest, Turnagain Arm to the south and the Chugach Mountains to the east.

As cities go, Anchorage is compact so one can live in a suburb and yet be down town in just a few minutes. Such was my own house. I lived in Lower Hillside on the lower slope of the Chugach Mountains.

Three miles up the hill from my house was Chugach State Park, encompassing seven hundred eighty eight square miles of spectacular wilderness. The Park Service long ago mapped out a series of trails in the closer regions of the park. Locals and visitors walked many of these trails. In winter the same trails were used for cross country skiing. Only authorized motorized vehicles were allowed within the park boundary.

Hagen and I planned a practice hike to limber up for the challenges of our big hike. Years ago, the local electric company strung power lines southeast between Anchorage and Girdwood. The trail, through Chugach Park and over the saddle of the mountain, to Indian Creek was called The Powerline Trail. About fifteen miles, it made a good hike with tough stretches up and over the saddle. From my house it would be a total of eighteen miles and we planned to spend one night camping on the trail.

We discussed our training hike during the week and packed our gear in preparation. It wasn't difficult. With only one night on the trail, we didn't need much food. We'd carry perishable items, like rolls and fresh fruit, which was impossible on a longer hike. Hagen insisted we pad our backpacks to the weight we would carry on the big hike. "Otherwise," he said, "it won't be realistic training."

I weighted my own pack with spare boots, heavy gauge plastic tarps, a couple of five pound bags of rice and four cans of beef stew. I finished with fifty-eight pounds, including my 30-06 rifle. I figured it would satisfy Hagen.

On Saturday morning, Hagen drove from Wasilla to my place. We drove to the southeastern end of the trail, Indian Creek. I parked my vehicle at a friend's house, then drove home in Hagen's Jeep and he parked it in the garage. We were all set.

Securing the house, we hefted our backpacks and rifles and set off toward the park entrance. We were hiking uphill and began breathing heavily and perspiring freely. As usual, I reached a point where my legs felt they couldn't take another step. I couldn't believe my load was only fifty-eight pounds. I reminded myself to throw away my obviously defective bathroom scale when Hagen called the first halt. Typically Hagen set the pace and was the one to decide when to start or stop. In some ways hiking with Hagen was akin to having a military drill instructor.

The first few miles were paved and I had to comment how quickly we could have covered the distance if we had driven. Of course that would not have served the purpose of toning our muscles for the big hike. To my utter relief we took a couple more breathers on the way to the park entrance.

Once we entered the park, there seemed much more logic

to walking. There were no more vehicles zipping by us. The trail was narrow and quite steep in places so that only an all-terrain vehicle could have negotiated. Further into the trail, where it crossed the ridge, only very specialized vehicles, such as those used by the utility company, could handle the steep grade.

It was eleven thirty, so we walked half a mile into the park, set down our packs in a sunny patch and prepared a lunch snack accompanied by a can of soda. While we picnicked, a few casual walkers passed by, but we were the only ones out for a serious hike. We were not yet on the Powerline Trail. We were on a well used trail which cut across the lower slope of Flat Top Mountain to join with the main trail two miles into the park.

We were still below tree line and surrounded by spruce trees and northern hemlock with occasional clumps of silver birch. In many places lower growth dwarf willows and alders crowded the trail. The hillside, with a western exposure, received a lot of sunshine and the trees grew large. Many were grotesquely twisted and stunted due to the weight of ice and snow in the winter time.

To the south, we could see the Kenai Peninsula, and, because we were familiar with the territory, could pinpoint Kenai fifty-five miles away. To the west, one hundred miles away, stood the Alaska Range—massive and gleaming with its year round snow cover. We had a clear view of Mt. Susitna, rising out of the vast Susitna plain forty four hundred feet. Anchorage was still out of sight behind the shoulder of the low ridge. We'd soon cross the ridge to join the main trail.

After our snack we crushed our soft drink cans, stuffed them in our backpack side pockets and set off again. Fifteen minutes later we crested the ridge and obtained a panoramic view of the city and Anchorage bowl. It was hazy, possibly a combination of traffic exhaust and airborne dust. Some days the city air quality was bad, especially on days when there was no wind. On the hillside, there was a hint of a breeze, the air was fresh and clean and we could smell the nearby vegetation.

Ten minutes more hiking brought us to the main trail and we had a view of it snaking up into the mountain pass. From here we could see and hear the waters of the south fork of Campbell Creek charging down its narrow rocky channel. On the other side of the creek were the power lines, strung on twin, wooden poles, following a line which converged with our trail two miles ahead.

Hiking up the gently rising trail took us above the tree line. The terrain around us was open except for occasional patches of dwarf willows and low growing alders. The trail underfoot was now rocky and more difficult to walk on. We were gaining altitude with every stride. As we did, the groundcover changed to arctic-like tundra with only low growing plants, coarse grasses, caribou moss, and white lichens.

Soon we passed a stake, placed by the Park Service, which read "twenty six hundred feet elevation." The north face of Flat Top Mountain, to our right, was still partly covered with rotting avalanche snow where the sun had not reached. We felt the air getting noticeably cooler as we climbed higher.

This was the third time we had hiked the Powerline Trail, so we were familiar with the area. We hoped to be over the ridge and to level ground before camping for the night.

The trail was nearly level as it made its way east into the valley across the lower slope of Suicide Peak and we made good headway. Our pace slowed however as we tackled the steeper trail leading up toward the ridge. The surface underfoot was rocky and we had to be careful where we placed our feet. The power lines, which had been getting closer, were overhead. Now we really were on The Powerline Trail.

We had taken short breathers but tenacious Hagen was in the lead. He had the bit between his teeth and wasn't about to stop yet. His stamina was at work. My thigh muscles were screaming with agony. He eventually stopped at a rocky outcrop half way up the steepest section. He must have been feeling the strain too, but typically, wouldn't have admitted it. This was not the last rest we took but eventually we hauled our aching and perspiring bodies onto the saddle of the ridge.

One minute we were scrambling upward almost on our hands and knees and the next minute we were on level ground. It was just as sudden as that.

Barely fifty yards ahead, the ground fell off steeply toward Indian Valley. We were on a very narrow crest of the saddle.

Thankfully, we put down our rifles and dropped our backpacks in a sunny spot. It was cool on the high ridge and a light wind made it feel much colder so we slipped into our jackets to avoid a chill. Hagen brought a small propane stove—as part of his ballast—so we decided to boil water for coffee. I had a couple of bananas and some granola bars which went nicely with the coffee.

Our view to the west from this high vantage point was unrestricted, framed by the bulk of Flat Top Mountain on the left and Wolverine Peak on the right. Spread out below was the City, and although we were too far away to make out detail, we could see the high-rise buildings of downtown silhouetted against the gleaming surface of Knik Arm. Forty miles away was Mt. Susitna, nicknamed Sleeping Lady, because of its distinctive profile. In the vast space between the Knik Arm and Sleeping Lady was the glistening, awesome, Big Susitna River pouring its glacial silt-laden water into Cook Inlet. The western view ended one hundred miles or so distant in the foothills of Tordrillo Mountains, parts of the Alaska Range. The mighty range marched southwest, eventually to form the islands of the Aleutian Chain. The view was captivating, yet made us feel diminutive and humble.

We soon felt the chill of the saddle, in spite of the bright sunshine. We didn't dally too long. Our snack finished, we packed away our things and moved on. Our muscles had tightened up a bit and it took a while to get moving. It was downhill all the way to our planned campsite, but walking downhill taxed different muscles and we knew it wouldn't be easy. It wasn't. The steep slopes were mostly loose shale. We both slipped and fell several times as our feet skidded out from under us. The rifles proved to be most cumbersome.

Anyway, we walked, slipped, and skidded our way a thousand feet down the steep zigzagging trail to the tree line where the footing was better. The trail leveled and began to traverse the north eastern slope of Suicide Peak, converging again with the power lines which plunged straight down the mountain side. The little used trail now became overgrown with grasses and low shrubs but the walking was much easier.

In the shadow of the mountain, the light faded. On the other side of the valley, the sun was still catching the mountain tops and the fields of snow were taking on rosy glows. It wouldn't be completely dark for a while, but we didn't want to wait too late before making camp. Walking for ten more minutes brought us to an open space which would serve our purpose well. We were always nervous about bears and preferred to have a reasonable field of view rather than be closed in by trees.

Our chosen site had a soft covering of moss and lichens which would help make our night more comfortable. We never carried a tent when we hiked. Instead, we had large ex-military ponchos which could be set up if necessary against the weather. They were really only nine feet square of waterproof material with a hood sewn in the middle and eyelets all around the outer edges. We needed a four and a half foot-long willow, or spruce bough, for a center pole, then the outer edges could be secured with cords and lightweight stakes. The ponchos proved to be effective as temporary shelters. We could see out under the edges to make sure there were no big hairy feet approaching. We liked that. I personally could think of nothing worse than being zipped up inside a tent unable to see what might be making a noise outside.

We set our ponchos close together, so we could easily talk to each other, and placed our ground sheets and sleeping bags underneath. Next, we collected fire wood, though we had a propane stove to do our cooking. What was camping without a camp fire? We soon had a cheerful fire burning within a small circle of stones. Using the stove, we heated a good sized pot of stew with chunks of Spam added for extra body. We settled back and ate this along with some crusty rolls. A kettle of water came to the boil for tea.

We enjoyed our evenings, depending on weather, by a campfire whenever we were hiking. It was great to relax after a strenuous day and feel aching muscles unwind. We felt comfortable in the wilderness aside from a little apprehension about bears. Frequently we had commented about feeling much safer in a remote place than in a city. It could be unsettling if, on darker nights, wondering what was just outside the circle of firelight. Occasionally, in the past, we had seen pairs of gleaming eyes, but it was just a curious fox or smaller animal.

As we sat and talked in the glow of the fire under a night sky, the conversation turned to the subject of our big hike. We had already decided to drop the tools on to the prospect area and discussed dropping smaller packages of supplies. The aircraft must not be overloaded. Items would have to be chosen carefully. We discussed the various options and decided we'd limit supplies to ten cans plus a sealed one gallon paint can containing a miscellany of snack items. For ease of handling, they'd be divided into three packages, sealed in plastic trash bags and then wrapped in tarpaulins. It would make them less attractive to bears or wolverines. We were sure the contents of the packages would survive the drop, especially landing on soft ground. Hagen suggested we tie surveyors tape to each package making them easier to locate on the ground. Good idea.

Hagen said it would be good to have an all terrain vehicle for easier traveling. We discussed this often and had already considered two second hand vehicles. In the end, we decided to reserve judgment until we hiked the trail at least once, then we'd have a better idea of vehicle requirements.

The proposed trail, scouted twice from the air, led forty miles along mountain ridges to a point north of our prospect area. Several low saddles along the trail, which from the air, looked wet and slippery. There were also some very steep slopes which might be beyond the capabilities of a certain type of vehicle. We wouldn't really know for sure until we had covered the ground on foot.

Occasionally we lapsed into silent reverie while gazing into the flames. I knew Hagen's thoughts at times like these because he would suddenly start telling me of some daring wilderness exploit he had read about. He read many books and had a fascination for the tales of the early explorers, trappers, prospectors, and mountain men. He envied them for the time in which they had lived, pitting themselves, with meager supplies, against the wild frontier. Their greatest weapons, their wits and the will to survive.

It was getting late and mosquitoes were trying to make a meal of us. We forgot our repellent. A quick trip to the bushes, brushing our teeth, cleaning our faces and hands, and we were ready to hit the sack.

Inside our poncho-tents, we suspended a piece of mosquito netting, carefully draped, to keep out all but the most tenacious insect. We kept our rifles by our sides in case anything larger than a mosquito came to bother us.

Our sleeping bags were good quality, double quilted with a filling of down and Holofill fiber rated to 20 degrees. They would roll into a compact bundle, yet had a high loft when fluffed up. After seeing the idea in an outdoor magazine, I purchased some soft flannel material and commissioned a friend to make liners for both sleeping bags. It was a touch of extra comfort and we adopted the habit of sleeping almost naked. We placed our clothes in a plastic trash bag to protect them from the dampness of the night.

With more than a few muttered curses we struggled out of our clothes, got into our sleeping bags, and arranged our mosquito netting. It wasn't the easiest of tasks under the confines of the poncho, but we were settled in a few minutes.

As usual, Hagen found something more to talk about. Though we were tired, this night was no exception. We talked about Hagen's plans. He worked for the Alyeska Pipeline Service Company. It was his ambition to own a machine shop, with metal machining, welding, and fabrication capabilities. We often discussed getting it started. What kind of work was available? It was no good hoping for an occasional odd job. There had to be some specialty or service no one else was providing. Also, it was important to have some core work to provide a steady income. Really we were brain storming and every time we did, we discovered something new to add to a list of possibilities. Hagen could get excited about this long-term planning for his future. Finally, during a lull, I drifted off to sleep. I left Hagen with his dreams.

In the bottom of the valley, Indian Creek, gorged by snow melt, rushed and tumbled its way to Turnagain Arm. Its soothing sound was all that penetrated the night.

I woke from a deep sleep and opened my eyes to green. It took a moment or two for me to realize where I was and that the sun was shining through the green material of the poncho. I had slept soundly, oblivious to lumpy ground under my sleeping bag.

There was a rattling of cookware and I realized Hagen was already up and about. Already? When I looked at my wrist watch, I was amazed to see it was eight-thirty. I had slept soundly for nine and a half hours.

I struggled out of the sleeping bag and into a tee shirt. Hagen had a cup of steaming coffee ready. He had been up just long enough to get a fire started and water boiling. He said, "A cow moose was browsing on the shrubs over there when I crawled out of my poncho. They aren't timid this close to civilization."

It was a lovely, cloudless morning. Warmth from the sunshine was already easing our muscles. The end of the trail was close, only about two and a half miles of easy going. We wanted to go to The Pines polka dancing that evening so we had no real reason to hurry. We lounged around, enjoyed the sunshine and had a good breakfast with more coffee. Slowly, we packed our things and broke camp.

We literally romped to the end of the trail. My Chevy was where we left it. There was no sign of my friends, so I scribbled a quick note and left it in a conspicuous place to let them know we arrived safely and had picked up the vehicle. Later we were back at my house. We got ready for dinner and an evening of polka dancing—like we needed the exercise!

## Chapter 5
## Polka Night

The Pines polka group started playing at around five PM Sunday evenings and continued with only short breaks until nine-thirty. Then a Country Western group took over and played until one-thirty AM.

Open every day except Monday, The Pines was a popular place in Anchorage. It was particularly well patronized Thursday, Friday, and Saturday evenings when the music was pure Country. Sunday nights were polka nights and a different clientele, family mostly, were attracted. Parents brought their children for the early part of the evening. The dance floor was about thirty-five by sixty feet, with a raised platform at one end for the musicians. The areas on each side were carpeted and there were tables and booths for four-hundred people. Through a side door was The Yellow Rose Restaurant where excellent priced meals, as prices in Alaska go, were served.

Hagen and I showered, dressed and set off, with both vehicles, at four-thirty. We went straight into the restaurant and ordered our meal and coffee. While we were awaiting our order some of our regular dance partners came in. Soon there was a crowd. We related our recent hiking experience and took a bit of a ribbing for not inviting some guests along.

We razzed them. "Shucks, we challenged, "You're mere females and couldn't have kept up with us super fit guys." Of course the challenge could not be taken lightly and we were promised company on some later hike. We were despicable. We never mentioned we were in training for an extensive hike. Still our secret.

Hagen and I routinely went to The Pines not only on Sundays, but on other nights of the week. Thanks to a lady who gave dancing lessons around town who finally, after much nagging, persuaded us to join. One thing led to another and we were drawn into aerobic classes too. In a short time we were involved, especially during winter months, and wondered how we ever managed without it. We became acquainted with many new friends. Our new friends introduced us to dancing at The Pines and other places around town. With lessons and practice, we were proficient dancers, and as a bonus, it helped us to keep fit.

Of course it was not lost that we were both eligible, single, and of considerable worth. Some opposite gender set their sights on us. We had used some of our new found fancy footwork to dodge when necessary but enjoyed close encounters.

On occasion, a relationship had been traumatic for one reason or another. We were not totally insensitive, and we had relied on each other for morale support and counseling through difficult times. What made Hagen and me such good friends was the fact we held no secrets from each other, yet we accepted each other with all our foibles.

Among our friends were, hikers, skiers, boaters, hunters, and fishermen. On weekends, eight or ten of us got together and enjoyed recreation such as cross country skiing with bonfire and picnic. We had a good time. We were introvert, in former times, but we were now borderline extrovert.

Everyone finished dinner and we went to the dance hall. Hagen and I selected our usual table in prime stalking territory. From our vantage point, we faced the music, watched the dance floor, and scouted the area for potential dance partners.

We didn't have to hunt much. We'd often be approached to "do this one with me." There were evenings when we hardly missed a dance. We had our favorite music and favorite partners, for certain dances, so it worked out well.

Neither Hagen nor I were big drinkers, but there was no cover charge at The Pines. They had to make enough to pay the band so we usually made our contribution by ordering a couple of beers. Besides, a drink on the table top staked your claim. On polka night the hall was not too crowded and most people had an established place to sit. The unwritten rule was to leave another person's regular place alone. Newcomers were usually invited to join one table or another until they settled in a group where they felt most comfortable.

The leader of the polka group called The Polka Dots, was piloting the band through a rousing number. We hardly had a chance to sit before we were yanked to our feet and onto the floor. I thought my feet and legs would not take any more punishment after this weekend. Of course they did. I was here for exercise. It didn't take long to loosen up and soon I was jinxing around in time to the music like everyone else.

The Polka Dots had a knack of throwing in a waltz now and then. To change the tempo, they played rousing numbers such as The Chicken which required some pretty ridiculous moves and gyrations. Then they required everyone to exchange partners for the next waltz or whatever. The kids enjoyed it and it was a good ice breaker for adults too, once one learned not to be self-conscious.

As I said, we had favorite tunes and favorite partners. We also had not-so-favorite partners no matter the tune. Hagen, who was particularly adept at categorizing people, had partners all sorted out. There were those who were a pleasure to dance with anytime. There were some to be avoided at all cost. There were good lookers, worth the sacrifice, even if they couldn't dance, and there were, unfortunately, some falling into the category of dogs. All this categorization, had it been revealed, would for sure, put Hagen and me into the much despised category of male chauvinist pigs. To a certain extent, it was a classic situation, eligible male hunting eligible female, eligible female hunting eligible male. Both determined to move so fast they were never caught.

We had developed a private rapport regarding the opposite sex, so if I spotted what I thought was a likely partner, I drew Hagen's attention to her. Hagen looked, and after some deliberation, made the terse assessment, "Worth a try" or "Forget it," or even "trollop, guttersnipe, stable-wench, hussy." Other times it was he who started it going and I who shattered his illusions with similar disparaging comments. Sometimes, there would happen by someone truly "awesome" which left us both speechless. Later, we maneuvered to be the first to ask her to dance. As we found out more about her, she ended categorized along with the others. We always kept these comments between ourselves, of course, lest we fall from that coveted title of eligible. The ladies, I'm sure, had their own categories for guys but we were never sure where we fit.

We gradually earned respect as good dancers and took credit for coaching many of the ladies in the room. We were still occasionally bombarded with requests to, "lead me through this one one more time." Sometimes it worked in our favor and sometimes it was just one of those dogs with a gleam in her eye. Above all, we tried to be chivalrous and sometimes suffered because of our largess. We had, on the other hand, developed friendships with many in the group.

Sunday night at The Pines was usually quite satisfying and tonight was no exception. The large room didn't get smoky on polka nights, like it did on the wilder Country Western evenings. That suited our desire for a healthy lifestyle.

On this evening, in spite of some protests from my legs, I hardly sat out a dance. Hagen and I both said our good night at ten o'clock and headed off to our respective homes and mundane stuff like working for a living.

## Chapter 6
### Flight to Gulkana

Living so far north, our days in the summer were long. A private pilot, without night certification, could fly almost around the clock. The FAA ruled daylight as, one hour before sunrise until one hour after sunset. It was now the last weekend in July and the days were long.

After an unavoidable two-week postponement, due to my business travel, we were at last making our flight to the prospect area to drop our supplies. It was a pleasant Friday evening without a cloud in sight as we set off. Our long flight took us through several mountain passes so I spent thirty minutes carefully preparing a flight plan. I phoned my flight plan in to the Flight Service Station and now activated it as we departed from Anchorage.

The baggage compartment behind the seats of the Cessna was full. There were three neatly wrapped packages of canned food, a bundle of tools, two sleeping bags, and ground sheets, and a little nylon bag of miscellaneous items. I was worried about the weight. But with enough fuel, plus reserve, it checked out marginally okay. One good thing, we had thousands and thousands of feet of runway for take off at Anchorage.

It was seven PM when I fired up the engine and obtained

clearance to taxi to runway 32. As we taxied, I called the FSS and opened my flight plan. Then, using as much runway it seemed as a Boeing 747, we took off and climbed steadily over Knik Arm to Point MacKenzie.

Our route would take us from Anchorage, north to Palmer and then northeast through Chickaloon Pass, Tahneta Pass, and across the flats to Gulkana. Throughout most of the journey we would keep Highway One, the Glenn Highway, in sight. The last few miles we cut across country before descending to the Gulkana airport.

It truly was a beautiful evening. All around us was a clear view of mountains and valleys. A hundred and fifty miles away Mt. McKinley and Mt. Foraker stood big, bold, and gleaming with snow. There really was no need for navigation as such because we were familiar with the terrain and we could see where we were and where we wanted to go.

First we headed north to Palmer at twenty five hundred feet. At Palmer, we turned northeast and climbed to our cruise altitude of ninety-five hundred feet. We were well below the flight path of larger aircraft which might approach Anchorage, but we kept a keen eye for small craft. The Cessna purred at twenty one hundred rpm and one hundred and four miles per hour. At this speed and altitude, the land below slipped by. Later, when we reached altitude, our progress, relative to the ground, would be less apparent.

This air corridor from Point MacKenzie to Palmer was busy, so we both kept a sharp look out for other aircraft and I made small course corrections as necessary. One speck in the sky turned out to be a large Bald Eagle. It zipped closely by at one hundred and four miles per hour.

Twenty minutes after takeoff, we had Wasilla in sight to the west and Palmer only minutes ahead. I touched base with the FSS to let the attendant know who we were and where we were going. I started a full power climb to our planned ninety-five hundred-foot altitude. Now this was the kind of flying I enjoyed immensely. It was wonderful to watch the view expand as altitude increased. We reached a point where we seemed to just hang there, up in the bright sunlight, with an awesome vista of mountains. The sun was behind us. At an angle, it left the deeper valleys etched in shadow.

Hagen was doing a bit of map-work as we flew along, not for navigation, but to identify visible roads, trails, streams, and lakes. It was useful to be familiar with the territory.

"Later," he mumbled, "we may find our way in for fishing at a remote lake. Maybe near the Glenn Highway."

Soon we were flying with the Matanuska River directly below and the well-traveled Glenn Highway to the left. To the right, however, was Chugach Mountain wilderness with no access roads whatsoever. From our lofty perch we could see valleys rarely, if ever, visited by anyone. We had often wished we could get in there to prospect, but short of hiring someone to take us by helicopter, there seemed no easy way.

It was wonderful. Exciting and spectacular country made our minds run wild. We could almost visualize gold nuggets mixed with gravel of the eroded river valleys. Hagen pointed to places we could hike, at some later date, and do a little prospecting, near the Glenn Highway.

We were making good progress. There wasn't even a ripple in the air as we flew over Chickaloon Pass. On our left was a six thousand foot peak and on the right an eight to ten-thousand-foot mass of mountains. Ahead, to the right, we could see the gleaming blue/white ice of the Matanuska Glacier rippling under peaks nine thousand feet high.

Below us, Matanuska River, spawned by the glacier, made its way down the valley. From our height, it looked quite tame. However, we knew it was a raging, silt-laden torrent making a very exciting raft trip for brave-hearted tourists visiting the State.

Ten minutes later we had a better view of the Matanuska Glacier as it stretched fifteen miles into the valley to its source near Finland Peak. During the summer, guided tours were taken over the lower reaches of the glacier. We had never done it ourselves, but it was popular with the summer visitors.

As we passed the glacier, our route took a slight turn to the left. We kept the Glenn Highway in sight. Gun-sight Mountain, so named because of the notch in its peak, rose massively to the left. Ahead we could see the deep cleft of Tahneta Pass.

Tahneta Pass is thirty nine hundred sixty-feet elevation. Flying fifty five hundred feet or so above the terrain, we could see considerably more detail. We passed Sheep Mountain and we could see white Dall sheep in the green upper meadows and on barren rocky slopes. To the north of the mountain, hidden from our view, was a valley with several feeder creeks which were actively mined for gold. On our return we intended to fly over the area and have a look. It was a shorter route but did not afford a view of the highway.

Soon we were over Eureka Summit and the Eureka Lodge at three thousand two hundred eighty nine feet elevation, according to the sectional chart. Over to our right was the gleaming expanse of Nelchina Glacier and, the northern tip of Tazlina Lake which is fed by Tazlina Glacier, itself still hidden by the shoulder of the mountains.

Around these two massive glaciers, the mountains, still snow covered in many places, rose to a height of eight thousand feet. The huge ice fields of Valhalla Mountain formed a wonderful gleaming backdrop. It was a spectacular panorama from Eureka Lodge but even more so from our high vantage point.

We had a magnificent view. In all directions, we saw mountains, each with its own shape and personality. The Amphitheater Mountains were to the north, the Wrangells to the east, the Chugach to the south, and the Talkeetna Range behind us. In this clear evening, even from this altitude, we were surveyors of an incredible wilderness area covering thousands of square miles.

The Wrangells, ahead, captivated us. They were a chain of spectacular mountains. Mt. Sanford, the most northerly, and with a classic pyramid shape, dominated at sixteen thousand two hundred thirty seven foot, snow-covered feet. Mt. Drum, Mt. Wrangell, a sleeping volcano, and Mt. Blackburn, lay off to the southeast. Finally the highest southerly peak was Mt. St. Elias. All stood dramatically on the eastern edge of the huge Copper River Valley.

I piloted our aircraft toward Gulkana. The Glenn Highway diverged to the right toward the wide spread community of Glennallen. Straight head, at a distance of twenty-five miles, was Gulkana Airport. Five minutes later I reduced the engine rpm, careful not to cold-shock the engine, and trimmed for a gradual descent toward the airport. We were on schedule, if not a little ahead of time, and would easily arrive before the FSS closed at nine thirty.

About two miles out, I contacted the FSS and informed the attendant of our intention to land. There was no reported traffic, but I switched on the landing lights—to make us more visible—and we checked to make sure all was secure for landing.

We passed over the Alaska oil pipeline, the Richardson Highway, and then turned onto final in line with the runway. The Cessna's tires squeaked on the Gulkana runway's asphalt, elevation fifteen hundred and seventy eight feet above sea level. We were just a few minutes earlier than estimated. As we taxied, I radioed the FSS to close out the flight plan and thanked the attendant. We would probably be his last customer for the night.

Hagen and I found a spot by the hanger to park the plane and our bodies for the night. We knew no one would mind us sleeping by the plane. It was quite common to just spread a ground sheet and sleeping bag under the plane, crawl in, curl up, and go to sleep.

I taxied the 150 to a vacant tie-down at the outer end of the apron, shut down the engine and systems, and pushed the plane into line with the anchor points. Though we planned to stay with the plane, it was always good policy to tie it down. There was no one there at the service center so we would have to wait until the morning to gas up.

After we stretched our legs by walking to the simple washrooms near the hanger, we unloaded our ground sheets and sleeping bags. It was iron rations tonight. All we had was a couple of ham and cheese sandwiches and cans of 7-UP.

We lounged on our sleeping bags for a few minutes to eat and drink but very soon realized every mosquito in Gulkana was homing in on our warm blood supply. There was nothing to do but spray on some repellent and slip into our sleeping bags for further protection. It was too early to settle for the night. Could we call it night? Not really. The sun shone, angled, over the north western horizon. Typical in Alaska were the nearly twenty-four hours of daylight and billions of pesky mosquitoes.

I pulled my sleeping bag up as far as possible to keep out light and persistent mosquitoes. Eventually, I went to sleep still sensing the hum and motion of my 150.

# Chapter 7
## Air Drop

Saturday morning was beautiful, sunny, and warm. After a good night's sleep, we got up early and packed everything and were ready to get on with our day. Unfortunately, the Service Center didn't open until eight o'clock on Saturdays, and we had to kick our heels for two hours. We had nothing left to eat or drink so we were gasping by the time the attendant arrived.

He did a good job though, and in ten minutes had a pot of coffee ready. He produced a package of supermarket donuts. In a small lounge, we drank our coffee and in the indoor facilities we freshened up a bit.

I pumped five gallons of fuel into the tanks of the Cessna. No need for more. It was only one hundred and twenty miles to Tok, our next stop. Besides, I didn't want the plane to be any heavier than necessary for landing and takeoff. We planned to land at Tok, have a meal, and then fly twelve miles to Tanacross. It was the only way to get a meal. There were no restaurants, or facilities, at Tanacross.

The Flight Service Station opened, so I walked over to check on weather and to file our flight plan to Tok. The whole area was under high pressure and no significant changes were expected. The fine weather would hold for several days.

We were tucked into the confines of the Cessna about nine o'clock. I cranked up the engine and taxied to the runway. Our takeoff was lengthy, but safe, and after climbing straight out, we commenced a very gentle climbing turn to the right and followed the Richardson Highway northward.

Five miles north of the airport we took a turn eastward to Tok. The Richardson Highway continued north toward Paxson, Delta Junction, and Fairbanks. We made a right turn to keep Highway One in sight. We climbed steadily toward ninety five hundred feet. Below, the Gulkana River and the Copper River joined, after draining the valleys of the Amphitheater Mountains, the Mentasta Mountains, and the Wrangells. Countless smaller streams flowed into the Copper River. To the southeast, about twenty-five miles away, Mt. Drum and Mt. Sanford stood pristine and majestic in clear air.

Our route was over sixty miles of level terrain, but we were heading straight for a deep notch called Mentasta Pass. Sinona Lodge, Slana, and Duffy's Tavern hunched along rivers and in valleys. They were all small isolated communities. From Slana, Nabesna road stretched south fifty miles toward Devils Mountain and Nabesna. Talk about remote! On the map, a mountain was marked near Nabesna, and called, intriguingly Gold Mountain—Hmmm!

The Cessna ran smoothly, the cabin temperature was comfortable, and we were moving along at our usual one hundred four miles per hour. Though I wasn't really checking, I had the impression we were riding a tail wind. In any case, it didn't take us long to reach the narrow pass entrance. Steep, impressive mountain sides shadowed the highway as it snaked back and forth through the narrow gorge. We maintained a straight flight path by cutting over the shoulders of the mountains to the left or right. Part way through, the valley widened and we had an excellent view of Mentasta Lake and the small community built along its southern shoreline.

Exiting the pass easterly, we followed the highway again

and Tok was about twenty miles straight ahead. I followed the highway, and descended gradually to the town. The sixteen hundred feet of gravel runway lay just to the right. It was parallel with the highway just before the intersection of the Glenn Highway and the Al-Can Highway. From the air the runway looked similar to the one at Wasilla, though trees grew a bit closer all around.

Tok landing-strip is uncontrolled but overseen by Northway, some distance to the southeast. I called Northway and advised we were about to land at Tok. We were on the down wind leg and looked around to see if there was any other traffic. I wasted no time in turning onto base and final. I kept it short and steep, to within a few feet above the gravel and flared down on contact. Hagen tensed up beside me but it was only the approach that was exciting, the touch down itself was smooth and gentle. I was proud of it. I maintained back pressure on the control column to hold the propeller well clear of the loose gravel and taxied back to the small tiedown area at the northern end of the runway. I kept the engine running for a few minutes to cool down gently while I radioed Northway to close out our flight plan.

We parked the plane alongside a blue and white Cessna 150 owned by an elderly resident we met on an earlier visit. He lived in a small cabin right by the tie down area. At seventy-three years of age, he was fit and healthy and was still flying. He once told us he flew across country to Whitehorse Yukon Territory several times a year to visit relatives. He enjoyed our visits and told us we were welcome to park the plane by his anytime. We pounded on the cabin door but there was no one at home. His old, beat up, Ford pickup was nowhere in sight, so we figured he must have been out on some errand. There was already a big pile of freshly split firewood stacked by the cabin. Maybe he was out harvesting in preparation for the winter. Although the cabin had electricity, he relied totally on his wood stove for heat.

Straight across the road was the rustic Tok Motel and restaurant. We headed there for breakfast. It was a bit late, being ten-thirty, but they said it was no problem. Ten minutes later we were putting away eggs, ham, hash-browns, toast, and coffee. We took care of the bill, noting it was almost double Anchorage prices. Oh well, Tok was, after all, on the famous Al-Can highway and it was still tourist season.

We walked two hundred yards along the road to the general store and purchased cans of Coke, bread rolls, cheese slices, and a packet of cookies. More Al-Can prices but at least we would not go hungry on the return trip. Back at the plane we conducted a walk around, checking everything, and were soon revving up at the end of the runway. I wanted to take advantage of every foot of runway because the plane was still heavy. It was a long rattling roll. I made sure we had plenty of speed before lifting off and used the extra momentum to carry us up and clear of the trees. Judging by Hagen's white knuckles, the trees were a little too close for his comfort.

"What are you doing?" he said between clinched teeth, "Paying me back for all those punishing hikes? When are you going to find long, smooth, asphalt runways?"

I just grinned.

Tanacross was only twelve miles away, so I held an altitude of only eight hundred feet as we made the short trip. We soon had the runway in sight and contacted Northway to announce our whereabouts and intentions. The wind favored the longest of the two runways— Hagen's kind— paved and forty-two hundred feet long. I piloted onto the downwind, looked around for other traffic, and set up for the landing. It was a beauty. No white knuckles this time.

There was an odd collection of aircraft parked around the hanger area, including an ancient DC-3. As we taxied, the Dakota was just starting its engines, with a cloud of blue smoke sizzling into the clear air. We puttered politely out of its way to the far side of the apron. We later learned from the attendant that the DC-3 was leased to the Bureau of Land Management and the people on board were counting sheep.

We located a station attendant and asked if we could fill, and leave the right-hand door with them for a few hours.

He said, "Help yourself. Leave the door inside the hanger. I leave at five so be back before then or it's mine for the night."

We assured him we would be back long before five. He didn't ask where we were going without our door. He was used to tight-lipped hunters and miners and tactfully didn't ask any questions. We had plenty of runway, but I put only a few gallons of fuel in the tanks. Just enough to keep us on the safe side for the round trip.

Now came the interesting part. Hagen took his seat fully rearward and strapped himself in good and tight. Then I wedged the cumbersome tool bundle by the side of his legs. When he was all set, I climbed into my seat, started the engine and taxied onto the runway. We learned from the first flight, with the door removed, and used foam rubber ear plugs to protect ourselves from the noise. The takeoff was long, and noisy, but we made it without the bundle interfering with the controls in any way. Hagen hugged it close for the takeoff but relaxed once we were settled at four thousand feet.

There was nothing to do but fly the sixty miles as directly as possible to the drop zone. With Hagen so confined, he really couldn't see very much. However, I picked up the trail head and kept it in sight all the way to Trib 1. I visualized just how far we were going to have to hike.

It took forty-five minutes to reach the area. I easily located the spot we had chosen for the drop. We intended, if possible, to explore three individual valleys, with tributaries flowing south into the Ladue River. On our map we named the valleys, from west to east: Trib 1, Trib 2 and Trib 3. We would be dropping the supplies by the side of Trib 1, between the stream and the ridge separating Trib 1 from Trib 2.

From the air, we could see the first stream, Trib 1, flowing down its narrow channel toward the Ladue River. Eventually, via the Ladue and White River, the stream contributed to the mighty Yukon River.

We planned everything, even marking the drop zone on an earlier photo. I piloted the plane in a wide, flat circle and lined up with the spot we had in mind. Meanwhile, Hagen untied the streamers of tape and eased the bundle outward into the slipstream. He could lean forward then, and monitor our approach. It was my intention to make a practice run, as we had done at Wasilla, and I coaxed the plane down to a hundred feet above the ground. "Not too low." I whispered to myself. As we reached the point where the land rose gently away from the swampy valley bottom, Hagen, almost casually this time, threw the bundle outward and downward. It was done so spontaneously and so smoothly I hardly realized the bundle had gone.

Hagen yelled; "Let's go!" I needed no urging.

I was already applying power and scrambling for altitude. Gliding down into a place where there is no runway seemed to go against my grain. Well! So much for a practice runs.

As we were going around, Hagen shouted, "It looked good. Let's not wait."

I agreed. The less low flying we did out here the better. Still, we had to go around again and drop the food packages.

In the few minutes it took to circle back into position, Hagen maneuvered all three packages to the doorway and held them there. He was getting pretty brave. He stuck his head out into the slipstream in an attempt to spot the streamers of the tool bundle. I concentrated on slowing the plane and bringing it toward the target. I spotted the streamers and aimed the plane just to the left of them. In case Hagen hadn't seen them, I shouted a countdown. "Three, two, one, go." Hagen hurled the packages and eased himself back into his seat, giving me a thumbs up sign. I got busy and made sure we climbed away safely.

The drop seemed to have been perfect. One more flight over the site revealed a cluster of packages, with streamers clearly visible, barely fifty yards apart.

Well! There was nothing more we could do about our

packages. They were now at the mercy of the wilderness and we would just have to wait until we hiked in to see how well they survived the impact.

Hagen now cranked his seat forward and made himself comfortable. I piloted the plane north, up the length of Trib 1. There, the valley formed a narrow cleft in the main hillside. We scouted around carefully and satisfied ourselves that no one had been there. At least there were no visible vehicle tracks of any kind. We went east, over the ridge to look at the Trib 2 Valley. Its character was very different to that of Trib 1. Trib 2 was slower and serpentine. The valley looked swampy in spots, and there was a sheen of water in places where it flooded. We had doubts about Trib 2 being a suitable area for our kind of mining.

The next valley, Trib 3, was much more promising. Its source was further away from the main trail, but it had steep, rocky overhanging bluffs. There were many places where the hillsides were eroded and shale slides went right down to the stream. It was obvious there was erosion and hydraulic action taking place and thus a good chance of finding placer gold.

As we flew south out of the valley, Hagen pointed to some vehicle tracks leading into the valley from the southeast. It was a faint trail but nevertheless, a trail. Someone had been there at some point in time not too long ago. Perhaps it was a hunting party, or like ourselves, someone prospecting for gold. We had no way of knowing. At least it was at the furthest point, six miles from where we intended to start prospecting.

Our view around the area completed, I pressed the plane up and over the highest point of the main trail along which we intended to hike. The map showed this point to be three thousand nine hundred fifty feet elevation and about equidistant from Trib 1 and Trib 2. It would mark the end of our hike along a defined trail, and to whichever of the tributaries we reached, it had to be through rough terrain onward.

I flew the plane parallel with the ridge and the trail to

allow Hagen to scout as much as possible. We paid particular attention to the low places on the saddles. We were interested in two aspects of these low points: first to see if they were wet enough to cause a problem, and; secondly to mark our map with potential sources of drinking water. As it turned out, our scouting payed off later.

Hagen was seemingly quite relaxed by the open doorway and was scribbling notes now and then on the edge of the map. This would all advance our knowledge of the area and better prepare us for the hike. A couple of times, at Hagen's request, I made right-hand circles around a point to give him a better view of a feature of particular interest.

Eventually we reached the last slope where the trail dropped steeply toward the Taylor Highway. We had seen all that we could and the next time would be from ground level and done the hard way. It was unbelievable that we were contemplating hiking those forty miles to the prospecting area. Forty miles on a flat map, may be half as much again on the winding trail, following contours of the land.

I pointed the nose of the 150 toward Tanacross and raised Northway on the radio to inform them of our intentions. They reported traffic for Tanacross but we were able to slip into the pattern and land with no delay. Our bombing run took two hours and ten minutes and we hit the planned target.

# Chapter 8
## Return to Anchorage

The Tanacross attendant cut us a sideways look as the 150 rolled to a stop. He refrained from asking questions. We reinstalled the right door and topped off the gas tanks. This time I filled the tanks to capacity because we were already fifty pounds lighter and were flying nonstop to Anchorage. During a walk around the plane, I checked the oil and found it was down a quart. The little Continental engine ran well but used oil.

When the plane was prepared, we sat off to one side and built ourselves sandwiches from the rolls and cheese we purchased earlier. We washed them down with Coke. We'd snack on the cookies later, and, on the spur of the moment, bought two cans of soft drink from a coin operated machine at the service center.

I spent a few minutes finishing our flight plan. Airborne, I'd call in to Northway. This was Hagen's kind of runway so as we taxied out, I handed control to him. He completed the run-up checks and with no traffic in sight, we lifted smoothly off the asphalt. Hagen set in a coordinated climbing turn while I raised Northway on the radio and relayed our flight plan.

Hagen was a fair pilot. He had taken flying lessons years

earlier, had soloed, but didn't follow through to get a license. Occasional practice in my 150 enabled him to improve his skill as a pilot. This takeoff and long flight gave him a good opportunity for practice. I just let him get on with it.

I was already glassing around at points of interest as Hagen leveled the plane at eight and a half thousand feet, cut back slowly on rpm to twenty one hundred, leaned the mixture, and trimmed the controls. There was no need for navigation, Hagen was flying visually straight for Mentasta Pass and then Gunsight Mountain which, still far away, was already visible.

I unfolded the Sectional Aeronautical Chart and occupied myself with identifying various features of the terrain below. Sectionals have a scale of 1 to 500,000 and are produced with surprisingly good detail. I began to identify twists and turns in the rivers and the highway, the confluence of streams with other streams, and the relative position of lakes. By this method I could tell exactly where we were at any given time. For practice, I noted our time between points of reference and computed our ground speed. Although our indicated speed was one hundred four mph, our actual speed, relative to the ground, was only ninety four mph. This wasn't significantly different to the eight mph head wind I used in my flight plan calculations, so there was no need to revise our plan. If the conditions remained steady, we would only be a few minutes late arriving at Wasilla.

When we reached Duffy's Tavern I dialed in 115.6 for the Gulkana VOR and let Hagen practice flying TO the beacon which radiated radio beams at five degree increments. Again on our left, loomed Mt. Sanford and Mt. Drum, snow-covered and glinting in the bright sunshine. It reminded me of a story I had read about the mountain. Apparently a plane, ferrying a load of gold for the World War II effort, disappeared and crashed into one of those mountains. It was never located. The plane, crew, and load of gold, was still on Mt. Sanford or Mt. Drum.

Strange too, we had never heard of anyone attempting to

climb sixteen thousand two hundred thirty seven feet Mt. Sanford. I supposed someone did at some time or other.

Ripples on a glassy surface of a lake below and to the left of our course, caught my eye. With the aid of the binoculars I saw a moose wading in the water. It was probably up to its belly and was grazing on the underwater plants. The ripples on the surface of the water gave it away. Of course it helped to have clear air, free from pollution.

Gulkana Airport was in sight and I radioed the FSS to let them know we were passing directly overhead to Wasilla according to our flight plan. A couple of minutes later the little flag built into the VOR instrument tripped over and displayed FROM, showing that we were now flying away from the beacon. I rotated the bezel to 229 degrees and Hagen made a slight turn to the right. He just had to keep the needle centered and it would lead us to Gunsight Mountain, which we could see over the nose of the plane. It was just instrument flying practice for Hagen.

The sun, almost straight ahead, was blazing in through the plexiglas windshield. We lowered the visors and wedged a map across to block out the sun so we wouldn't get sunburned. The map was positioned so our view was unrestricted. When we made this flight the previous year, we got sunburned and ended up terribly red faced in Wasilla.

Hagen maintained a straight course all the way to Eureka Summit and then scooted to the right of our VOR radial, and flew us over the placer mining area on the north side of Sheep Mountain. Many minor streams converged in the valley to become Sheep Creek which then ran southwest into the Matanuska River.

We were amazed to see the activity on the streams and the main creek. "It looks like there is a mining operation every few hundred yards." I told Hagen. "I can see the creeks flowing muddy brown from each mining site."

"Settling ponds are not required at most placer mines." Hagen commented. "I suppose the Matanuska is so heavily silt-laden, the state had to make exceptions."

We left the western end of the valley and saw the Glenn Highway snaking its way alongside the river. On our left, was the great expanse of the Matanuska Glacier with its clearly visible streaks of silt and rocks being carried along by the ice flow. The end of the glacier looked more like massive gravel-works from this height but we knew there was still ice under much of the gravel.

Now we flew westward with the mountain range rearing up close on our right. We were at only eight and a half thousand feet, and those mountains looked massive. Hagen drew my attention to several flocks of mountain sheep, white, against the high green alpine meadows. Ahead we saw the wide green expanse of the Matanuska Valley farming belt.

Wasilla lay to the west. Hagen piloted the plane to the right and we slipped over the last few ridges by Jonesville Mine, and still fifteen miles out, throttled back a little and initiated our descent. Hagen stayed in control while I contacted Palmer to tell them of our arrival and close out our flight plan. Hagen turned onto the down wind leg for our landing at Wasilla.

There was no traffic in sight so Hagen stayed in control, went through the landing checks and set up our approach from the southwest. He plunked the 150 down solidly, rather than gracefully, onto the gravel runway, bounced, and had the chance to land a second time. Our flight had lasted three hours and thirty-eight minutes. About fifteen minutes more than initially planned.

I was satisfied with our trip. As we taxied toward the house, I glanced at Hagen and gave him a thumbs up. He was not smiling, probably because he was still wondering about his one approach and two landings. I could tell he enjoyed the experience of being pilot in command for the flight home.

Thirty minutes later we were relaxing with a light snack and a cup of tea. We accomplished our objective to place supplies and tools on the prospect area and we knew more about our next long hike.

At six o'clock, we gassed up, then flew to Anchorage. Hagen's vehicle was at my house so he'd stay the night and go straight to work in town in the morning. Today was Sunday and Polka night at The Pines. We had time to get ready, hit the restaurant for a well-deserved meal, and get a little exercise on the dance floor.

## Chapter 9
## The Drive

With the heavy supplies out at Trib 1 there was less to carry on our backs. It still required some careful planning to make sure we carried enough to sustain us on the long trail, supplies to give our diet a bit of variety, and enough for the trip out.

It was our plan to hike part way in, and, at an appropriate place, stash makings of a few good meals. This would make our packs lighter for the remainder of the trip in and ensure we had sustenance on our return.

We had nearly two weeks to round up goods for our backpacks. I had my own list prepared during the first week and compared it with Hagen's list to check for duplications. We remained self sufficient as far as breakfast and snacks were concerned but pooled our canned food, dehydrated meals and cookware.

In the remaining days I assembled items on my list and laid them out on the floor of my spare room. Looking at the list had been bad enough, but as the pile grew, it seemed like an awful lot to squeeze into one backpack.

My backpack had upper and lower main compartments, two large pockets on the back, and two pockets on each side. Arranging everything for the long hike was a challenge. The art was to try to arrange things in the order they would be needed. The side pockets were the most accessible and held water bottles and a miscellany of smaller objects which might be required any time.

The large lower compartment held my spare clothes well wrapped in a plastic trash bag. The upper compartment held my food and cooking utensils. Ground sheet and sleeping bag, tightly rolled and in their own waterproof bags, were attached with bungee cords to the bottom of the pack. A small, long handled, axe, aluminum shovel, and gold pan were strapped on the outside by more bungee cords.

A strong leather belt carried a ten round 30-06 ammo pouch, a folding buck knife, and a small soft pack for miscellaneous items. The 30-06 rifle had to be carried by hand since there was no way to satisfactorily sling it on a shoulder or strap it to the pack.

With everything packed, the load weighed sixty seven pounds. Not too bad and it would be lighter when I cached my share of the meals on the trail. I wondered what we would have done if we had not dropped the supplies in advance. I suppose we would have managed somehow but it would have been a mundane diet of dehyd and more dehyd.

We'd leave Thursday. I had my pack ready by Tuesday evening. I did think of a couple of extras but they fitted into the side pockets and it was ready. No! The sun can burn down something awful at the higher altitudes and I had forgotten the sun block number fifteen lotion. A quick trip to the store. I was ready.

Well! The big day at last. I walked a few miles for exercise and retired early. I slept well and the alarm woke me at six. Quickly I piled out of bed, showered, and was dressed.

The Chevy Blazer was ready and my hiking gear was in the back. I secured the house and set off toward Wasilla, fifty-five miles away. There was heavy business traffic on Lake Otis, Tudor Road, and Muldoon and I was delayed a few times.

Once I reached the Glenn Highway, it was better. I was a few minutes late at The Pantry restaurant. Hagen was seated in a booth and already into the morning newspaper.

We were eager to get going so we wasted little time ordering breakfast and eating. "No time for fooling around with politics today, Hagen," I said wiping my chin. At eight thirty we left the restaurant and drove our vehicles to his house.

Earlier we had agreed to take my Chevy Blazer on this trip. Hagen was a bit nervous about leaving his nearly-new Jeep unguarded in some remote place. My older vehicle would be of lesser concern, though I didn't want anything to happen to it either. We took protective measures of course and loaded some dirty old tarps and a camouflage net into the back of the vehicle. We had purchased them at a surplus sale in Anchorage and they would serve to cover and disguise the Blazer when we stashed it near the trail entrance.

Hagen's pack seemed very heavy as I helped load it in the back of the Blazer and I suspected he had not been careful measuring quantity of the contents. He was always apt to "throw in a couple of these just in case" which, of course, added to the weight. I knew however, Hagen would not be hungry on the trail even if it meant carrying extra weight. Hagen's Weatherby 30-06 rifle was the last item to be tucked in the back before closing the tailgate and cranking up the window.

The day brightened considerably and looked like a pleasant one for our journey. We locked the garage and set off. Just to the east of Wasilla, a newly paved road cut through to the Glenn Highway near Palmer, the initial part of our route. At the Palmer crossroads, we turned left. It was the Glenn Highway all the way to Tok Junction on the Al-Can Highway, three hundred and sixty miles away.

We knew the Glenn Highway closely and with dry roads would be able to make good time. A main problem was getting past slower moving vehicles but our knowledge of the road and its twists and turns helped.

We made good headway. When we reached the lookout point for the Matanuska Glacier, I pulled off the highway to stretch our legs and enjoy the view. The glacier, lay gleaming blue in the sunlight. In the glacial area were mounds of glacial moraine which, as we had observed from the air, resembled a huge gravel works.

We were just about to leave when an Alaska Tours bus loaded with Japanese tourists arrived. It was amusing to see one older lady snapping photos even as she descended the steps of the bus. She ran off about six frames before she even reached the guard rail. I guess she thought the glacier might melt before she had a chance to capture it on film. They smiled and bowed politely to us as we retreated to our vehicle in the face of their excited chatter.

The terrain in the Matanuska region is rugged and the road twists and turns through unique scenic beauty. Shades of green foliage, devil's club, tall grasses and berry-laden bushes converged in harmony. The hillside areas and roadside embankments were covered with large patches of red fireweed, blue lupines, wild roses, and a variety of smaller flowers. For the few summer months, Alaska was a garden spot, beautiful and rich, not as supposed a land of ice and snow.

An hour later we topped the rise at Eureka Lodge where we stopped for a cup of coffee. Good coffee too. Eureka Lodge is three thousand feet above sea level. Across the road was a gravel runway, and beyond, miles of open country and patches of skinny black spruce trees. In the distance, to the south, glaciers, ice fields, and massive snow covered peaks mirrored the sun and formed a spectacular backdrop.

Keeping our stop brief, we paid the tab—unbelievably only fifty cents a cup—visited the men's room and were soon on our way.

The sixty-mile stretch of highway from Eureka to Glennallen had few points of interest, unless one cared to count the frost heaves, but we made good time. Ahead of us was the snow covered, sparkling bulk of Mt. Drum with Mt. Sanford a little further away. It looked like clear weather ahead. As we closed in on Glennallen, I drove closer to the fifty-five mph speed limit because I knew this stretch of road was patrolled by State Troopers. We had no problems. By one o'clock we reached outskirts of the town and pulled into the gas station.

We topped off both gas tanks. From this point we had one hundred and seventy miles to travel to the entrance of the trail. We planned to have a good dinner at the Tok Motel restaurant and decided a light snack now would not be unreasonable.

The little restaurant next to the gas station was bustling with tourists. We ordered cheeseburgers and fries and then waited. They were delicious. "I doubt we'll need a big meal at Tok."

Hagen retorted, "Wanna bet?"

He was probably right. If we hiked to the crest of the ridge, once we reached the trail, we wouldn't want to prepare a meal at such a late hour.

Only two miles further the Glenn intersected with the Richardson highway and we turned north. Twelve miles north we turned east again at the Tok cutoff. The sign by the junction read, "Highway 1, Tok and Canada." It was just an extension of the Glenn Highway.

As we turned east, the road took a mile long down grade and, in very spectacular fashion, crossed the wide gravel bar expanse of the Gulkana River. A short sharp climb and the road hugged the northern embankment of the Copper River. We had occasional spectacular glimpses of the Copper, but for most of the time, the river was obscured by trees.

We made good time despite more frost heaves. There was little traffic in either direction to give any problems. Many vehicles were motor homes and most carried license plates from the Canadian provinces or the Lower 48. This was the main artery from Anchorage to the Al-Can and toward the Canadian border and Whitehorse in the Yukon.

One hour after lunch break saw us passing the Nabesna junction at Slana. Just as we had seen from the air, Slana was only a few cabins strung out along a short stretch of highway. Another hour placed us almost at the entrance to Mentasta Pass. Streams and tributaries seemed to come from all directions and for a few miles we drove over a wide river bed. Alaska might run out of oil one day, but it would never run out of gravel.

Soon the scenery changed from spectacular to very spectacular. Great cliffs of rock pressed close to the road, mountains literally hung threateningly over us. The road, through the pass, swooped and zoomed dramatically, lending to some exhilarating travel. Ten miles into the mountain pass, the valley widened out and Mentasta Lake, with a few cabins along her edge, slowed us down. Though pretty, we concluded it must be one mean place to live the winter. A road side sign stated: Elevation 2,300 feet. I was surprised it wasn't higher but supposed that was why the surrounding mountains seemed so large. They rose to seven thousand feet.

Twenty miles further, we left the pass in a northerly direction. As we came down the final grade, we could see the flat lands of the Tok and Tetlin area spread out ahead of us. The Tok River was off to the right side, and though we couldn't see it, the Tanana River cut its way in a northwesterly direction across the flats. In the distance, maybe fifty miles away, we could already see the low line of ridges which was our destination.

We passed through the western-most reaches of the Alaska range and Tok lay twenty-five miles ahead. The road bed was so full of frost heaves we slowed to save our livers. There were delays due to road work and we had to wait at one place for a pilot vehicle. We passed a large motor home suffering a broken rear spring. He was limping along at ten miles an hour and would eventually make it to Tok.

At Tok we made the State Troopers office our first stop. It was just good sense to let someone know where we would be for the next ten days. As Hagen said, "So that someone could bury our bones one day." The Sergeant on duty however, took it quite seriously and said he wished more people would do the same. He also, quite seriously, asked if we were packing guns to protect ourselves and we confirmed we had 30-06s. We promised faithfully to check in upon our return.

He said, "We'll set the dogs loose, just to locate your bones, if you're not back in twelve days."

This essential business completed, we drove half a mile to the Tok Motel and restaurant. It was too early to eat dinner but nevertheless, we ordered a fair plate with Chicken Fried steak, mashed potatoes, mixed vegetables, a side salad and coffee. It cost a lot—Anchorage times two—but, as Hagen pointed out, it would be our last civilized meal for ten days.

Finished with our meal, we eagerly headed east along what could now truly be called, The Al-Can Highway. The huge bridge over the Tanana River teased us. One more mile brought us to Tetlin Junction where the Taylor Highway struck north. The Junction was poorly marked and could be missed easily. The only indication was a small sign reading Chicken and Dawson. Also, if you knew about it, there was an old cabin adorned with dozens of Moose and Caribou antlers occupying the corner lot.

We knew the Taylor Highway consisted of one hundred and eighty miles of gravel leading past Mt. Fairplay, Chicken, and Jack Wade to a border crossing called Boundary. Heading east into Canada the road arrived at the Yukon River where there was a ferry to Dawson City, site of the legendary Bonanza Creek gold strike.

I drove the Taylor Highway years earlier and remembered it as awesome. For many miles the road ran along a high ridge. It seemed like driving on top of the world. The highest point was four thousand five hundred feet and kind of lonely. It was definitely not a place for a breakdown. It could take a day or two to get a tow truck and it would cost a bundle.

Our trail began only twelve miles up the gravel highway. From aerial observation, we knew there was a large gravel pit almost opposite the end of the trail. Easy to locate, it took only fifteen minutes to reach the place. Road crews had obviously used this gravel pit as a source of material for the road bed. Now the large level area carved out of the bluff, yet partly overgrown, more than anything else was used as overnight layover for motorhomes and the like. There were no vehicles parked now, which was good, because we would prefer not to be observed entering the trail.

I drove the Chevy just a little way in from the road bed and switched off the engine. We climbed out and were instantly aware of the silence, a silence broken only by the pinging of the Chevys' hot exhaust system as it cooled.

A bit stiff, we walked down the road fifty yards and searched for the entrance to the trail. It was overgrown and difficult to locate, yet right where we expected. The road had been graded recently and there was a fresh berm of gravel across the entrance as well as a heavy growth of alders. We were pleased to see there was no evidence of any kind of vehicle entering the trail recently.

We pushed our way through the dusty shrubbery and soon found ourselves in a partially overgrown clearing a hundred yards across and surrounded by large trees. It must, at once, have been the staging point for the mining company that carved the seventy-five-mile trail into the wilderness. Powerful Caterpillar tractors typically followed each other in what was known as a Cat Train with each Cat hauling sleds of heavy equipment for exploratory drilling. No one seemed to know what they discovered. There wasn't a big operation out there now.

Before exploring any further, we needed to get our own vehicle off the road and out of sight. It was important at this stage not to draw attention to our presence. I walked back to the Chevy and drove it down to the trail entrance. With Hagen directing, I eased the vehicle over the berm and through the bushes into the clearing. We both then set about carefully rearranging the gravel berm to obliterate any tire tracks. Not a moment too soon, we ducked back into the bushes as a vehicle rattled from the direction of Chicken. It passed by at fifty miles an hour and left us choking in a cloud of fine talcum-like dust. The dust rooster-tailed from the vehicle and settled over us. We both let out a favorite expletive and Hagen commented: "That's civilization for you." We dusted ourselves off and headed for the Chevy.

Now we had to find out how far we could drive in along the trail to stash the vehicle. We already knew it couldn't be too far. First there was a low swampy area and then a very steep climb which would be a challenge to anything but an all-terrain vehicle.

We were first struck with size of the trees. They were much larger than we expected and they had more of a Christmas tree shape, unlike most of the trees this far north. They would form an excellent cover for the Chevy.

We took our rifles and walked down the trail to scout. The further we went, the more soggy it was. After a couple a hundred yards, apparently we would take a big risk bringing the vehicle even this far along the trail. Ahead the area looked more like a pond than anything else and several acres on each side were spongy bog. It would be difficult to traverse on foot and impossible to drive over.

Well that settled that. We looked for a place to hide the vehicle without leaving much of a trail. That meant staying on higher, drier ground near the trail entrance.

We retraced our steps up and checked out an area on the north side of the clearing. Finally we agreed on a place fifty yards from the trail head. It was well hidden from the highway, and from the trail. Our camouflaged vehicle would be invisible from the air under the wide spread tree branches.

I drove the Chevy to the edge of the clearing. With Hagen guiding me to make sure I didn't flatten anything or hang up on a boulder, I eased the Blazer between the trees. Finally Hagen guided me under the wide spread branches of a tree and I stopped the vehicle and switched off the engine.

Satisfied with the location, we hauled all of our gear out and went to change into hiking clothes. We left our travel clothes neatly inside the vehicle ready for our return. I had decided to wear corduroy pants, a tee shirt, long sleeved wool shirt, socks, and hiking boots. I donned my strong leather belt with its various pouches. My down filled vest was tucked conveniently into the top of my backpack.

Hagen favored blue jeans, wool shirt, and his short Levi jacket along with socks and well used hiking boots. He also had leather belt to which he attached a black leather sheath which held a large Bowie-type knife. He tended to use this mean looking weapon for all sorts of things including spreading squeeze cheese on his rye bread at meal times.

When we were dressed, we covered the Chevy with tarps and camouflage net. By the time we finished it looked more like a lump of rock than a product of General Motors. A couple of days and the grass and shrubs would spring back up and the vehicles' location would be secret from all but ourselves.

We wrapped our keys, and wallets in a zip-lock bag and buried them just by the base of a nearby tree. They too, would be safe. We had no use for driving license, money or credit cards where we were going.

With a lot of grunting and groaning, we hefted the heavy packs onto our backs and picked up our rifles. We were ready to hit the trail and start our most challenging hike.

# Chapter 10
## Up to the Ridge

Ten minutes on the trail, we got into trouble. It was a good thing we didn't try to drive the Chevy through the wet area. It turned out to be much more swampy than we thought.

We tried to stay close to the original trail, but it was just too flooded. We back tracked a little way and went to the left, but we were on hummocks of marsh grass floating in water. Obviously, we were not going to get through easily.

We boldly took our pants off and, looking quite comical, walked through the water-filled tracks of the trail itself. The clean water with underlying mud, was preferable to cloying muck. A hundred yards of wet area and the deepest puddles came to our knees. We sloshed along snorting at our funny remarks and made it to where the water gave way to higher ground and walking was easier.

This was the starting point for the North Fork of the Ladue River. The area we passed through was the fringe of a catch-basin for runoff from the surrounding hills. Recent rain would have left the area more flooded than usual and we must have caught it at a bad time

Thankfully, as the trail heaved itself clear of the swamp, the ground became dry again. The trees were larger, well established, and anchored in firm ground.

We lowered our back packs to the ground, cleaned ourselves up and got dressed. It was a messy start to the hike. We were fortunate the only damage was wet socks and boots. We managed to get through without losing our footing and dumping ourselves and our packs in the water—a disastrous beginning. We dried the insides of our boots, rung the water from our socks and then delved into the side pockets of our packs for dry socks. They say: "Great minds think alike." We both produced socks enclosed neatly in zip-log plastic bags.

It took us twenty minutes to get ready to set off up the steep trail. We began to see visible evidence of caterpillar-tracked machinery employed in trail breaking. Here and there, large boulders were displaced to one side or the other. In one place, a huge rock had gouged a furrow of several yards before rolling to one side. The rock was easily two cubic yards in size.

The cat train had obviously picked its way between the larger trees and followed a gentle slope. However, tree trunks six inches in diameter had been pushed over and crushed into the dirt. Some timbers were still visible and sound in spite of the years. In Alaska, the cold preserves timber and there are fewer wood devouring insects, prevalent in other climates. Deep scars on the land and vegetation, made by the cat train, would remain for many, many years.

It was a steep climb and with our packs at their heaviest we were both being severely tested. Hagen, of course, was ahead and setting the pace while I brought up the rear. I could tell he was struggling too. We just had to take a break now and then to catch our breath and give our aching legs a rest.

The first rise topped out at three thousand eight hundred feet, according to our map. The trail, however, followed a long extension of the ridge and along many terraces. The cat train operators, had taken every advantage and had gone from terrace to terrace but resorted to winching sleds up steeper sections.

We were surprised at the lush foliage. A rich variety of trees and shrubs pressed in on the trail from each side. There were patches of silver birch, aspen, northern larch, spruce, and poplar. We noticed this great diversity from the air and saw something of a pattern. Usually, it resulted from forest fires. One species of tree was wiped out and cleared the way for another to gain predominance. The changing elevation played a role too. Spruce and larch at higher levels, silver birch and aspen at lower levels. Slopes with southern exposures developed different growths than northern facing slopes. Where there were no large trees, dwarf willows, or alders, formed a tangled barrier. The trail was partially overgrown in places and tough, wiry, knee-high shrubs snagged our legs as we walked. A touch of brilliant color, added by the tall red fireweed, stood among patches of small wild flowers and wild roses. We were fortunate to have the trail to follow even if it was showing signs of returning to nature.

Hagen and I staggered under the weight of our heavy loads. During a frequent break, I said, "I don't think we can make it to the crest of that ridge before night."

"There is nothing to stop us from making camp right here." Hagen wiped his face. "I prefer to be above the tree line away from mosquitoes, thousands of mosquitos."

We pressed on, took short breaks, and followed steeper sections. We took little comfort in knowing this was probably the hardest part of the hike. I groaned, "My pack is so heavy. We must not have limbered up that last hike."

"What about all that dancing at the Pines? Didn't it help?" Hagen offered.

As we climbed progressively higher, we gained a better view of our surroundings. We could see the Taylor Highway wending its way north. Mt. Fairplay rose above the surrounding ridges. Immediately to the north of our ridge, a wide green valley formed the drainage for the West Fork of the Dennison River. To the south, lay the North Fork of the Ladue River. Over a low ridge, lay Tetlin Flats reaching to the Alaska Range.

The sun was slanted toward the horizon and we realized we'd have to call a halt soon. We didn't want to set up camp too late. Ahead we could see the trees thinning so we decided we would hike just a little further and then make camp, no matter what.

Our legs seized up after every short rest but we gritted our teeth and staggered on for about thirty minutes. Unanimously we decided enough was enough. We reached a small terrace in the long slope and were just above most of the trees. The trail climbed on ahead through lower growth with only occasional stunted trees in sight.

"This is what we've been looking for." Hagen said as he struggled out of his pack. "At last, a place to camp above the mosquitos. And look at the view."

I sat on my pack and replied, "We didn't run into bears hiking, so maybe we won't, camping in the open." I suppose we still felt there could be a bear behind every tree or around each bend in the trail.

Scouting around, we found a level grassy patch just south of the trail. It was perfect for our camp. Scavenging around the immediate area, we brought together firewood for the evening and next morning. Hagen circled some small stones and I started a neat little fire within the circle.

The light was fading and as the sun went down, the northwest sky turned a bright orange, striped with shades of red and brown. We were bathed in an orange glow while the valleys below were in shadowy grays. At this time of the year, and at this latitude, it wouldn't get completely dark. The eerie twilight would remain until the sun rose again very early in the morning.

A couple of years earlier, Hagen cut and assembled some light weight stainless steel tubing, formed it into a four-legged frame for hanging a kettle. The lengths of rods were held together at the ends by rings. I quickly named it the Quadrapod and the name stuck.

We now suspended a kettle of water from the Quadrapod and while it was heating, we laid out our ground sheets and sleeping bags. It felt cool now. We had ceased our strenuous activity so we felt a need to slip on our down vests. Our boots were still damp from the soaking in the swamp so we hung them on tripods of willow branches and put on our running shoes—instant relief.

The water was boiling so we threw loose tea straight into the kettle, set it aside to brew for a couple of minutes, and stoked the fire with larger pieces of wood. Then we settled back to snack on squeeze cheese and crackers. They washed down well with hot, sweet, tea.

The twilight deepened and stars shimmered in the clear night sky. Except for the crackling and popping of our fire, there was absolute silence here on our ridge.

It was eleven o'clock by the time we were ready to turn in with the back packs close by and our rifles placed handily. We snuggled into our sleeping bags, naked except for our shorts and tee shirts. We talked as warmth crept through our bodies and our muscles relaxed. Eventually we drifted into silence and into our own dreams. The untended fire sputtered and died. A complete calm fell over our remote campsite.

# Chapter 11
## Day One on the Trail

When we awoke the next morning, valleys below were shrouded with thin layers of mist. Our ridge rose in clear air. Everything, including our sleeping bags, was damp with heavy dew, and the air was nippy. We draped ground sheets over bushes and laid the sleeping bags on them so the air could circulate around. Hopefully, they would dry by the time we had to roll them up and start hiking.

We started a small fire in the hearth and freshened up before preparing breakfast. Hagen found a patch of beautiful, juicy, blueberries nearby. We picked a few and I added them to my bowl of breakfast cereal. Hagen, rummaged around in his conglomeration of rye bread, cheese spread, and trail-mix and added the blueberries. The instant coffee tasted great, as did most foods, out in the open air.

Breakfast finished, we packed our gear, and damp sleeping bags, made sure the fire was extinguished, checked our rifles, and resumed our climb up the trail. The ground was firm and the clearly defined trail was fringed with low growth.

We progressed upward for half an hour before we came to a steep slope of loose shale and rocks. The tracks left by the cat train were clearly visible a hundred and fifty feet or so directly up the slope. It was going to be a tough one for us to climb. With a lot of cursing, straining, scrabbling for foot holds, we dragged ourselves upward under the weight of our backpacks. Gasping for breath, we hauled our shaking bodies onto the bald, flat top of the ridge.

We had walked almost three and half miles and climbed two thousand feet since we left the Chevy. Our map showed this first unnamed high point was three thousand eight hundred and six feet elevation.

What a wonderful view. The mist was burning off in the valleys below and we had a clear view to the south, west, and north. To the east our view was limited to a few miles but only by Mt. Son. Our hike would take us over Mt. Son. We had a new perspective of the trail ahead. Just because we were up on the ridge didn't mean it was going to be easy all the way. We could now see the trail dropping about five hundred feet before wending its way to Mt. Son's top. It was a dimension of the trail that wasn't clear, even from the aircraft during our scouting flights.

With such a good view of the trail ahead we used the binoculars to scan every yard. From this spot, the tops of the ridges looked dusted with white lime. It was the white of the Caribou moss on our hiking trail. This tiny plant, with its white coral-like leaves, grew tenaciously on the high, dry, tundra surfaces. They prospered where life was too tough for larger plants.

For a short distance, the trail dropped gradually into the trees. Walking along the ridge, we probably wouldn't run into swamp like the ones we first encountered. The ground fell off quite abruptly on both sides.

After a short breather, we pushed on along the trail toward the saddle. The slope was gentle on this eastern flank and walking was the easiest we had experienced thus far. The surface was covered with tough wiry mountain heather. A low growth of blueberry and alpine bearberry, sported crimson leaves. The tracks of the Caterpillar tractors and runners of the heavily laden sleds had left long gouges in the surface and caused severe erosion. We had to be careful where we placed our feet. We didn't need twisted ankles. We'd have a real problem walking out again.

The sun was well up now and a little breeze blew from the southwest. It was time for sun block on our faces or we'd be peeling by tomorrow. The tube was handily in a side pocket of my pack so Hagen accessed it without me putting down my pack. Hagen and I both got sunburned easily, yet tanned well after the initial exposure. Suitably protected, we continued our way.

After the down hill stretch, the trail roller-coasted gently for a couple of miles but not enough to make it hard work. It was a bit soft for a few hundred yards, at the lowest point, but not really wet. We were surprised again to find large straight trees. Some of them were about eighteen inches thick at the base and they were fifty to sixty feet tall. We were used to seeing skinny arctic black spruce everywhere. These larger conifers made us feel humble.

We were a bit tense among the trees because of limited visibility. This was new territory. We were further from civilization, and these realities worked on our minds and told us it just had to be bear country. Maybe it was, maybe it wasn't. We saw no sign of bears, but we took no chances and carried our rifles at the ready across our chests. We hoped we'd never have to use them.

About mid-morning we called a halt. We were just beginning to climb up the western flank of Mt. Son and it was already much steeper than we had anticipated. We didn't go to the bother of making a fire, but settled for a drink of Koolaid. A granola bar and some dried fruit filled the need for a snack.

It was a relief to put our rifles down, shuck off the backpack, and flex our tired shoulders. Thinking ahead, we spread our damp sleeping bags out over the shrubs to further dry. I was having problems with a couple of pressure points in my boots and if I wasn't careful, I would have to deal with blisters. During the break I changed my socks, hoping it might improve things. I found in the past that it made a big difference. Even a little wrinkle in a sweaty sock could be irritating enough to promote blistering.

Twenty-five yards south, the trail passed an outcrop of rock looking out toward the west. We were spellbound. The view over the west fork of Ladue was magnificent. The valley was beautiful and green. Trees hugged the serpentine river. Mature tamarack and spruce trees grew further back where land was better able to drain. The rest of the valley looked decidedly soggy and we knew it would be impossible to walk on except in the winter when it was frozen.

The slopes leading up the ridge, supported a growth of trees. At two thousand foot elevations, the larger growth gave way to dwarf willows. Fingers of trees occasionally made it to higher levels where there was a saddle in the long ridge.

With Hagen's pocket camera, we took a photo of each other posing as mountain-men against the wilderness. We hoisted our backpacks and rifles and set off again along the steepening trail.

Our leg muscles tightened up during the half hour rest, and it took a while to loosen up again. With tenacious Hagen for company, I was the only one to verbalize on this fact. Hagen only grimaced, set his already angular jaw more firmly and lengthened his stride. I satisfied my own ego by telling myself this was just a front. In truth, his legs were killing him, and on top of that, he was probably ravenously hungry.

The trail steepened sharply and we found ourselves laboring up the flank of an intermediate peak, unnamed on our map. Several difficult, steep, but short, stretches of shale left us gasping for breath. We took a rest after each. We noted the deep scars on the slopes, and in particular, at the upper lip of each slope where the sleds had been winched with brute force. No small wonder the trail was clearly defined after so many years.

Slowly, we made our way upward and finally reached a

bald plateau a couple of hundred yards square. By a quirk of nature, there was a small stand of stunted northern larch trees at the eastern edge. It was a good place to stop for lunch and a real rest. There was a pretty little spot in some large rocks and shrubs. It would have made a beautiful campsite but it was too early for that. Perhaps it would be convenient on the way out.

It was one PM and we had put six miles of trail behind us since our overnight camp. We were satisfied with our progress and agreed we deserved a rest.

Thanks to the nearby spruce trees, there was no shortage of fire wood. Ten minutes later, we had a kettle of water on the boil for hot tea, dehyd beef and vegie stew. I had a couple of crushed, but palatable, French bread rolls, to compliment the stew. Hagen had a couple of apples for dessert. These were luxuries available only because it was our first day on the trail. From now on we would not be able to enjoy fresh food.

After our meal, we removed our hiking boots and stretched out on the ground to rest. There is nothing that beats a patch of soft spongy, dry, tundra for comfort; forget the berry stains on our clothes. Despite his aggressive hiking fashion, Hagen relished a siesta after lunch. For myself, I took every chance I could get, thus it wasn't long before we were both sawing wood.

We dozed for an hour before rousing and even then we continued to lounge. Finally, we spurred ourselves into action, rounded up our gear, made sure the fire was out and moved—stiffly—from the plateau.

After a moderately steep downgrade, the trail leveled and continued along the spine of the ridge, thus providing excellent views to the north and south. The view prevailed for a couple of miles until we reached the lowest part of the saddle where the dwarf willows, were more prolific and—so much for dwarf—about eight feet high, crowded the trail.

Refreshed from our rest we held a very good pace along this easy section. Soon, we climbed the western flank of Mt.

Son and found it increasingly difficult. These steep sections, it seemed, were always much steeper with greater changes of elevation than we estimated from our flyover.

The slope up the west side of Mt. Son was something else. The average slope was perhaps one foot up in three feet forward, but there were steeper slopes with shale and rock slides to be negotiated. Here was a prime example of what happened once the surface was broken by vehicles and the thin veneer of protective vegetation destroyed. It's obvious there was little or no natural erosion until the surface was scarred by the cat train. The finer materials simply crumbled and washed downhill exposing larger rocks and boulders.

We slaved doggedly up these rock and shale slides with frequent stops to get our breath. I had to think how we flew over the trail these same miles in only a few minutes. Whichever way one looked at it, this was the hard way.

I was struggling upward, commiserating with myself, my head down, when I almost bumped into Hagen. He was frozen in mid-stride. He grabbed my arm to prevent me from falling back down the slope and pointed uphill with his rifle.

At first he made no sound but just worked his jaws. He squeezed my arm tighter and finally gasped, "Is that a bear?"

I looked in the same direction but at first couldn't see anything. My eyes, pretty good at a distance, while Hagen's were myopic, finally made out the object. It was only a dark patch of rock vaguely bear-shaped, framed by lighter shrubs. At a hundred yards, with a bit of imagination, it did look like a bear. Hagen was not fully convinced until we climbed closer and the bear turned into just a harmless rock.

One thing became patently clear because of this incident. We realized we were spending far too much time watching where we placed our feet and not enough time looking where we were going. Thinking about it, I couldn't remember looking behind often either. We heard stories about unseen, curious bears following hikers. A bear could have been behind us most of the way and we wouldn't have noticed, maybe until it was too late.

We vowed to be more watchful lest we stumble upon a real bear and not just a bear shaped rock. Knock on wood, we hadn't seen evidence of bears during the hike, but that didn't mean they weren't around.

We briefly rested on the bear, then struggled up the trail and eventually, conquering one last especially mean talus slope, staggered onto the top of Mt. Son. We reached the first really high point on the trail, three thousand eight hundred and twenty two feet above sea level. It didn't seem like an Alaska mountain, but our hike was influenced more by change in elevation than actual elevation above sea level.

This one sported a level area the size of a couple of football fields. We were disgusted to find a hundred or so, rusting fifty-gallon fuel drums on the northern side of the plateau's low shrubs. They could only have been left there by the mining company or possibly by the Bureau of Land Management. We were appalled to think they were allowed to leave their trash behind. Here we were, not even leaving a sardine can behind, while they, with all their mighty money and machines, left this mess. Sadly, over the years, we heard of many similar cases.

We settled on a spot by the eastern edge of the plateau, away from this mess, for our rest period. Two major summits had been conquered during one day of our hike, and yet it was still too early to consider quitting for the day. We decided to consult our maps and photos to see what lay ahead. To the east of Mt. Son the trail fell gently and then meandered along a high ridge about three miles before rising to a minor peak. Following this, it dropped sharply to what we believed to be a lowest saddle on the trail. We didn't want to get caught in a low, mosquito-infested place for the night, so we decided to hike to a knoll about three miles away. This would be our camp for our second night on the trail. By the time we reached that point, we would have covered sixteen miles or so. Pretty good for one day and actually much better than we expected.

Our rest on Mt. Son was short and we were soon making

our way down hill and along the ridge. This was a great stretch to hike and we made very good time, almost romping up to the rocky knoll.

We set our backpacks down and looked for a suitable campsite. Hagen suggested a spot on the south side of the knoll, a level area surrounded by stunted bearberry shrubs. It ended abruptly at a cliff which dropped vertically for about twenty feet. From the ledge, we had a marvelous view over the Ladue Valley and to the east. In places we could see the continuation of the trail.

Neighbors! Among the jumbled rocks at the foot of the cliff was a family of collared pika squirrels. Other than a few small finches and some spruce hens, the pika were the only other wildlife we'd seen. At least it was wildlife we could enjoy watching and not be afraid of an attack.

While they entertained us with their antics, we noticed a threatening buildup of clouds to the southwest. "Hagen." I said. "They weren't evident earlier in the day. Maybe we're in for a change of weather."

He headed back to our gear, mumbling. "We'd better get prepared just in case it rains during the night."

With a bit of searching we came up with enough firewood for the evening and for the next morning. We also found a couple of longer willows to use as center poles for our ponchos. We got everything set up while a small fire was reduced to cooking embers. Also, we stored some firewood under the shelter of our ponchos so we would have less of a problem the next morning if it rained.

We cooked up a good mess of beef stew and beans, throwing in some extra dehyd vegetables. With a hard day of hiking behind us, we ate, talked, and dangled our bare feet over the edge of the rock cliff. It was great. Hagen found that the squirrels loved the peas from the stew. We enjoyed watching them dart out and retrieve the ones we pitched. Chattering between themselves and with tails twitching, dash back to the entrance of their burrow. It became a game they seemed to catch on to very quickly. Since there was no name on the map for this knoll, we decided to call it, Squirrel Peak and later marked it so on our map

We made a kettle of Hagen's favorite blend of tea and relaxed as early twilight crept across the valley. Clouds slipped across the sky and hid the sunset. A slight breeze chilled the approaching night.

We took care of our ablutions after dinner and then sat by the stoked fire until we were too tired to talk anymore. There were no mosquitoes to worry about at this elevation. All was well. At ten o'clock, we rolled our tired bodies into our sleeping bags and lapsed into silence. I remember thinking about my throbbing feet when deep sleep overtook me.

## Chapter 12
### Wet Saturday

In the morning, we were awakened with rain drops spattering on the ponchos. Our foresight, and the ponchos, kept us, our gear, and firewood dry. We looked out on a miserable, dripping wet world.

We awkwardly pulled on clothes while still under shelter of the ponchos. Our Gortex jackets were essential before venturing out. The sky was grey and there were few signs of clearing. At the moment it wasn't raining hard, more like a damp mist mixed with rain. Visibility was less than three hundred yards.

There was no sign of our neighbors, the Pika squirrels, in the lower apartments this morning. Obviously they peeped out earlier, saw the miserable weather, and decided to go back to bed.

Smarter than us, Hagen mused. "Rain. Harrumph, must be a week end. There's no reason to hang around here. Let's get breakfast out of the way and pull up stakes."

About eight o'clock, we were on our way for the second full day of hiking. The downgrade gave us an easy start, though it was a bit slick in places and we had to be careful. We made good time and it wasn't long before we were down on the saddle and walking again in giant trees. The trail began to twist and turn to avoid the larger stands.

Drizzle turned to downpour just as we reached the lowest part of the saddle. The mushy surface of the trail began to puddle with water. We hurriedly donned our ponchos and huddled by the trail to rest and wait it out. The Gortex jackets were very good but they couldn't stand heavy, prolonged rain. Eventually water sneaked in around our collars and soaked our clothes. The hooded ponchos, on the other hand, spread out to protect us, our backpacks, and our rifles.

The rain showed signs of tapering off and the sky became a little lighter after about thirty minutes. While sheltering, we mulled over what to do about drinking water. We knew from our maps and aerial surveys there was a gully close to the south side of this low point, maybe a potential source of water. The foliage was sopping wet. It was the worst time to go plunging off the trail. But we needed to find water. There was no way we could consider tackling the next stretch of high ridges without an adequate supply.

Shortly after the rain stopped, we removed our ponchos and stashed our backpacks under the trees. Then, taking our rifles, the nylon day pack, and our water containers, we set off downhill.

We were soon soaked and we were feeling uncomfortable floundering through a tangle of dripping vegetation. A hundred yards further, to our complete surprise, the shrubs and grasses ended abruptly and we found ourselves in a grove of northern poplars. A canopy of spreading branches and very little growth underfoot. It was a bit like walking through a cultivated orchard of giant trees. They had smooth trunks twelve inches in diameter and the limbs started branching out eight feet above us. The orchard was a striking contrast to the rough growth in these parts. We had never seen such a plantation anywhere.

We walked easily under that great canopy and we soon found a substantial stream cutting across in a southeasterly direction. Our water supply was barely a quarter mile away from the main trail. The stream appeared a little swollen by the rain, but the water still ran clear. Parallel with the stream was a clearly defined bear trail. There were no fresh tracks, but it served to remind us there were animals larger than our neighbors on Squirrel Peak.

We wasted no time in filling our containers, including a collapsible two-gallon jug. Just to be safe, from now on we would boil any of the water before it was consumed.

With ten quarts of water, we set off up through the orchard. We felt pretty safe because we could see a hundred yards or so in any direction. However, the safe feeling was shaken. Large bones, obviously caribou bones, were scattered around on the ground. Wolves or a bear, had a feast on the carcass. The meaty bones had not been there very long either. We felt apprehensive about standing in the middle of what was left of a carnivore dinner. If this was a grizzly kill, we were asking for trouble because it would definitely be viewed as a violation of territory. Without wasting time, we got out of there, heavily loaded or not. This time, I might add, we remembered to look behind us.

We got ourselves a fresh soaking by the time we reached the backpacks. At least now we had a water supply and knew where we could get some on future trips. Later we identified it as Poplar Hollow on our trail map.

Making special note of a large tree by the trail, we stashed the food and water in reserve for our return trip. We also left a small collection of trail trash to pick up on our return—good planning that kept our load to the minimum for the rest of the trip.

Well, rain had stopped, but we were feeling wet and miserable. An hour and a half of our morning had been consumed by seeking shelter and water. We decided we should get dry before continuing.

We soon had a good sized fire going right there in the middle of the trail and some of our wet things strung up to dry. Off to one side we started a smaller cooking fire and boiled enough water for coffee and to make Koolaid for later.

By the time we hit the trail again, an hour and half later, we felt much more comfortable. We still had spare dry clothes. We couldn't take the risk of traveling without a spare set of clothes to fall back upon if necessary.

The rain held off and we walked quickly for a distance. The trail steepened and we slowly started a long climb toward the next high ridge. We had deposited food packs but noticed the additional weight of water we carried. For a couple of miles it wasn't too bad but then the trail steepened abruptly. This was the kind of slope the cat train winched itself to the top. We found ourselves scrambling, clawing, and cursing the last two hundred feet of boulder strewn slopes. The rifles were a real pain, but we wouldn't have been caught without them.

We finally made it gasping for breath to the top. This was the steepest slope we had yet encountered. It would pose a definite challenge to any kind of vehicle we might decide to use in the future. In spite of our condition, we chuckled when Hagen pointed out we were both "steaming" in the moist air.

The sky threatened, so we opted for another break. It was, we figured, past our lunch time, so we broke out iron rations of crackers, sardines, granola bars, and trail mix. It went down well with freshly mixed Koolaid. We rinsed the sardine cans with water before flattening them and sealing them in a Ziplock bag. We didn't want the bears smelling sardines and following us along the trail.

The weather seemed to improve even in the short time we rested, but it was still pretty damp as we hit the trail again. The trail followed the high seesaw ridge for ten miles. It was refreshing to hike faster, without stopping for the view. We needed to catch up on time and couldn't afford the luxury of stopping anyway. This part of the trail was without landmarks. It was difficult to tell where we were and the lack of scenery encouraged us to just keep plodding along. We forced the march and managed long periods without a break.

By late afternoon we approached the north western end of

the Trib 1 Valley. We confirmed that fact when the trail passed over a little rocky knoll and we got a better view of our surroundings. About four miles away, to the southeast, we could see the rocky cliff which stood prominently about half way down the eastern side of the Trib 1 Valley. The high point of the trail, cut in a wide loop, was about five miles away. We had to get there before descending into the valley.

It was a steady uphill climb, but we reached the highest point on the entire trail in two hours. The map gave this smoothly rounded dome an altitude of three thousand nine hundred fifty feet. Even on this cloudy, misty day, the high point afforded a panoramic view and we could identify all three valleys; Trib 1, Trib 2, and Trib 3, and the ridges separating them.

Evening—we were damp clear through. We had done enough for one day. There was no firewood on this high tundra peak. We decided to hike a little way along the ridge which separated Trib 1 from Trib 2 to find a suitable camp site for the night. We could see heavier growth not too far away.

This was now our first hike off the main trail. The crest of the ridge was high dry tundra with a thin veneer of low growth clutching tenaciously to the firm shale surface. It was easy underfoot so we made good time to a place where a few stunted and winter blasted spruce would be fuel for our campfire.

It's called dry tundra, but it wasn't too dry at the present. For our campsite, we picked a shale-covered, drained area with minimum vegetation. It might not be quite so comfortable, but it would be drier to move around on. As usual, we made starting a fire the number one priority, and then set up our ponchos for shelter and placed our ground sheets and sleeping bags inside. Only then did we set about preparing an evening meal and a large pot of coffee.

It was an evening of rest and recuperation within the circle of a welcome fire. We rotated our damp clothes on a line within the range of the heat until they were dry then made an early retreat to our sleeping bags. It wasn't really dark, but there was not much to be seen except cloudy skies. A mist shrouded Trib 1 Valley just below our camp.

## Chapter 13
## Trib 1 at Last

Sunday dawned bright through a thin, high cloud overcast. Great blankets of mist swirled in the valleys. It was cool, damp, with a hint of a breeze. The promise of fine weather was in the air and we hoped for a warm day.

I stretched and suggested to Hagen, "Let's make our way down to the supplies and find a good place to make camp." Whatever we did now might be temporary. Permanency depended on results of our prospecting.

We started a small fire and had a good breakfast which included beautiful juicy blueberries collected around our campsite. We rounded off the meal with a second pot of coffee and were ready to go by nine o'clock.

Hagen shrugged into his pack and said, "We can set up camp near Trib 1. Exciting being this close, huh? Our planning and preparations, and hard work, are about to pay off. Let's go."

We knew everything could look very different from the sky so we studied our maps and photos and identified a few land marks. I pointed, "Let's stick to our original plan and take advantage of the bald ridge, maybe three miles. Close to the southern end, we can take the shortest route."

Hagen concentrated on a photo. "Looks to be really steep."

"You lead out." I said, "We can do it. Trib 1, here we come."

As we walked, the weather brightened and the sun came out. It was instantly more pleasant. The spine of the ridge sloped gradually and finally supported a growth of willows and stunted spruce trees. The ridge seemed to lengthen but we eventually made to the point where we needed to strike off to the airdrop site.

We took a minute's breather and eagerly drank the last of our morning's Koolaid. "Hagen, let's dump the water." I suggested. "We'll be near the stream soon. It'll lighten our load too." With lighter backpacks, we set off down the slope. At first it wasn't too bad. We began beating our way through increasingly dense foliage. It was pretty too, with a southwestern sunny exposure, the upper hillside was covered with a plethora of wild flowers, fireweed, arctic forget-me-nots, alpine azalea, and rounded clumps of flowering moss campion.

As the slope steepened, we came to eroded rock and shale slides, difficult to negotiate. With our packs and rifles hampering us, we were hard pressed to keep our footing.

It was eight hundred feet from the top of the ridge down to the level of the stream and was hard going most of the way. We just picked our way slowly and carefully downward. We both lost our footing several times which resulted in scrapes and bruises. The butt of my 30-06 picked up more scars on this stretch than at any other time so far.

Finally the slope eased and we could hear the soft, babbling sound of Trib I as it made its way south. Fifty yards of dense willow and alder thickets, some ferocious thorny devils club, and we found ourselves on the bank of the stream, twenty yards away. We emerged to the north of our supplies.

Trib 1 at last. We slapped each other on the arm, then just stood, lost in individual thoughts. The stream was larger at this point than it seemed from the air. We figured it averaged eight feet wide and ten inches deep. Trib 1 was sparkling clear and coursed along at a healthy clip over a clean bed of sand and gravel. The gravel and earthen banks were four feet high. Since the stream followed a serpentine course, there were wider and deeper pools in the bends—sandbars, on the inside of the bends, undercut eroded banks on the outside. The stream was, we thought, just large enough to support grayling or trout.

We walked alongside the stream just a little way and found a place which we thought would make a good campsite. At least it would do for the time being until we had explored the valley. The site was elevated a few feet above the stream. It afforded an extensive view out over the Ladue Valley and was also far enough away from the stream for the sound of running water to be muted.

Dumping our packs, we took only one rifle and set off in search of the supplies. We had an aerial photo marked with an X. I guess Hagen spotted the bright streamers first, but there they were, all four packages. The packages had landed on firm ground. Externally, they seemed to have survived. The spade and pickax had cut through the burlap once again, but that didn't matter, the burlap was only intended to keep the contents in a manageable bundle.

We lugged the packages up to the campsite and set about unwrapping everything. It was a bit like Christmas, except we were careful to untie the bundles and save every bit of string, bungee cord, burlap, and tarpaulin. Out here we never knew when, or for what, these items might come in handy. We were really delighted to find the only damage was dents in a couple of the cans. Again, there was too much bow in the saw.

Now we had plenty of subsistence for the next few days. Also, we had our tools if we decided to get into serious digging during our prospecting.

Looking at our food stock reminded us it was lunch time. An energetic morning left us ready for a snack and a kettle of tea. We gathered stones, made ourselves a hearth, and soon had a small cooking fire cheerily crackling.

Browsing around, with a mug of tea and a granola bar in hand, Hagen commented. "It might be best if we keep our gear out of sight of any planes. After all we really didn't want to advertise our presence any more than was necessary."

I agreed. After some discussion, we decided not to build anything more than our ponchos until we scouted upstream. If we found a likely place to prospect, we'd want to relocate our camp.

After a little while, we tidied up our bits and pieces and with only our rifles, aluminum shovel, gold pans, snack and drink in the nylon day-pack, we set off to explore upstream.

At first, we held to the east side of the stream. There was no suitable place to cross without getting our feet wet. For the first half mile, the stream ran through open land. There were patches of aspens and spruce ahead on rising land. The drainage was obviously a little better there.

Sure enough, we found a small escarpment where the stream tumbled down a series of rapids. It was quite a picturesque spot and we figured there might be connection between it and the sheer cliff close to our right. Perhaps at some time there had been a landslide which formed the cliff and the escarpment.

In any case, the terrain was entirely different at the escarpment. Downstream the land was swampy with dry land close by the stream where drainage was better. Upstream the land had the appearance of a meadow, well drained, with a variety of silver and paper birch, aspens, and large spruce.

Here also the stream was not so deeply eroded and the bed was a mixture of larger rocks and gravel. There was a total absence of algae on any of the rocks and pebbles. It was as if they had been scrubbed clean.

It was a good place to dabble with the gold pans to see if there was any color. We looked into the clear water. No obvious flakes or even nuggets leaping out at us—too much to hope for—so we dug down and found fine gravel and sand. It was an interesting colorful blend of material, well laced with black sand, flakes of mica, and shards of quartz, always a good sign. After reaching a depth of eighteen inches we started washing the material in our gold pans and worked to get down to the heavier fines.

Hagen was the first to find gold. In his second pan he had a couple of tiny flakes mixed with heavy black sand remaining in his pan. The flakes were large enough to pick up on a damp finger. I did and peered at them with a ten power glass I had in my waist pouch. It was gold all right, rounded edges, soft in appearance under the glass and coppery, rather than yellow, in color.

Granted, the flakes were small, and small flakes, like these, were in many Alaska streams, if one looked hard enough. However, we were excited because we had only scratched the surface and panned just a very small sampling of material to find these flakes.

It was an exciting find too, because we had invested so much in; research, exploration by air, and hiking to reach this place. It would have been disappointing to have come away completely empty. Well, this was after all, just the start. There was still two miles of stream to go and undoubtedly it would get more interesting as we moved north to the end of the valley.

On the premise that "every little bit helps," I placed the two tiny flakes of Trib 1 gold in a vial inside a little Ziplock bag.

We tried to stay close to the course of the stream, but in some places it was impossible to penetrate dense thickets of alders and willows which flourished on the well drained land. The growth and the shelter provided by the patches of trees, sure seemed like just the place for a bear to call home. We went carefully, but not quietly. Give them a chance to go the other way, was my motto.

There were animal trails in all directions, but we suspected many of these were made by caribou passing this way during annual migration. It didn't help our feelings very much, however, and we were still apprehensive.

We had only traveled a quarter mile further when we spotted a pile of fresh moose droppings. This was moose country too, and we could expect to see one or two. In the wilds however, they were not so used to people and they were apt to lope away long before you got very close to them.

The ridge to the west of Trib 1 was much lower and rounded than the one to the east. Being much larger in surface area, it formed the major collection area for the tributary. Our aerial surveys showed several minor tributaries converged into a main stream, fishhook shaped at the head of the valley. The area was almost completely forested and we recognized stands of large Sitka spruce.

As we walked further north, the way became more difficult. The valley sides were closing in on us and the trees and undergrowth combined to block our path. Where there were no trees, low growing shrubs and brambles snagged our clothing and scratched our legs.

The ground underfoot changed and we came across jumbled out crops of bedrock. Now this was a bit more like it. Gold could not percolate downward, if there was bed rock. There was a good chance of finding gold at bedrock, if we could get down to the low points. All the signs were in our favor. There was color in the stream and now some bedrock. Perhaps a good combination.

The trees on our side of the stream were large, and one, about a foot in diameter, had bridged the stream. Although the stream was now small, we used the bridge and crossed over to clearer ground. We found ourselves on a level area, a gravel bar, on the inside a sharp bend in the stream. The outer side of the bend was formed by a low protruding shelf of bedrock.

The stream made a turn here. Its convergence was with bedrock on the east side while the west bank was a crescent filled with an overburden of materials washed from higher elevations. It was the kind of entrapment area prospectors dream about. It made us shiver with excitement at having found, so early in our search, a place with such appeal.

We rested for a few minutes, caught our breath, and contemplated the scene before us. On the west side, upstream, there was a steady incline caused by the low ridge and a cascading stream, over, and between, a jumble of large boulders and the protruding ledges of bedrock. We would explore a little more upstream later, but this level place by the bend was the most promising.

We came prepared to stake out several forty acre claims if necessary, but all we needed was a one acre parcel of productive claim. Anything more would be just a buffer zone around the active work area. This place by the bedrock might just turn out to be the "glory hole" we needed and we wouldn't have to search further.

Satisfied, we decided to carry out sampling as close to the bedrock as possible. There were a few fissures in the cliff and these formed V shaped miniature valleys which were filled with gravel and silt. These niches formed a natural entrapment for any heavier metals and would be an excellent place to start. Unfortunately, it entailed our wading around in the cold water and we had no rubber boots. Well, no one promised it would be easy on the last frontier. We took off our socks put our boots back on, rolled up our pant legs and, grimacing, waded into the cold water.

Hurriedly, not to prolong the agony, we dug down below water line, filled our gold pans with material and tipped the material into an accumulation on the bank. We repeated this several times at different places until we had quite a collection to work. After five minutes of hopping around on cold aching feet we put our socks and boots back on again and set about some serious panning. We each took small amounts of the material and, finding a convenient place by the stream, where we didn't have to wade the icy cold, began panning, separating heavy and light material.

Gold is about seven times heavier, size for size, than any common rock, gravel, or sand, it's not too difficult to wash away the lighter materials and keep the heavier material in the pan. The difficulty is to separate the heavier sands from the even heavier gold. Panning technique is important. We had panned for gold many times and were proficient so we speedily got down to having just a little heavy material remaining in the bottom of our pans.

Wahoo! Lots of fine gold dust, along with flakes gleamed in the bottom of our pans. Every sample we panned yielded a more fine gold and eventually one flake, rounded enough to be classified as a nugget. We began jumping around screaming and laughing with excitement. Underneath Hagen's usually stern veneer, he was exuberant and it took just a hint of success—or was it gold —to open the flood gates. He was ecstatic. I think he was willing, almost, to hike out and in again just to get a sluice box. A sluice box is what we needed to process this material in any quantity.

We panned through the first pile, and more, over the next few hours and were rewarded with quite a bit of gold. It tinkled in the vial by the time we finished. We found a good place to start. How lucky can you get?

It was great! In some ways it took the wind out of our sails and made us rethink what to do next. We hiked out here intending to explore at least Trib 1 and Trib 2, maybe even Trib 3 and then make a choice. However, here we were already, with what looked like a very promising claim. Could we expect to find better? We doubted it but, after some deliberation, we decided to have a look around anyway.

Hagen said, "I don't think I will be really satisfied until we look over Trib 2 and Trib 3. Knowing we didn't while we had the chance would give me nightmares." There are stories of people still finding thumb-sized nuggets in areas and we would hate to have missed the chance. Besides, we were curious about tracks we saw leading to Trib 3.

While we had a snack and a drink of Koolaid, we mulled over our options and decided to hike from our camp around the southern tip of a ridge and take a look at Trib 2. Next we'd move our camp to the Glory Hole and spend a couple of days taking samples. We'd attempt to find out how widely the gold was spread. Next we'd hike from the Trib 1

Valley to the main trail by the shortest route. It might be a very difficult climb but we were determined to give it a try. It would cut a lot of miles off our journey. Once at the main trail we would hike to a point nearest to Trib 3 and see if it was practical to hike down into the valley. It would keep us busy for a week. Finally, of course, we would hike back out along the main trail.

It was five o'clock so we decided to return downstream to our camp. Back at camp, we started our cooking fire and had a good meal of canned beef and vegetable stew and lots of hot tea. We were careful to boil the water before using it. Being careful of the water was good. Planning ahead, we boiled enough water to prepare Koolaid for our hike to Trib 2.

There weren't too many mosquitoes yet, so we had a good refreshing wash in the cold stream water. Pity however, we had to spoil it by applying insect repellent when the beasts decided to emerge.

Gauging the weather, we figured it wasn't going to rain but we decided to set up our ponchos and stow all of our gear for the night. At the very least, we'd benefit from having the mosquito netting. This done we sat around the campfire, sipped hot tea and talked until we saw the stars come out. Only then did we prepare to retire for the night. Finally, snuggled in our sleeping bags, our tired bodies relaxed.

We must have been thirty-five miles from the nearest human being during our first night at Trib 1 and the only sound was the muted babbling passing of the stream a few yards away. I fell asleep thinking about how lucky we had been to find color on our first attempt. I'm sure Hagen was already spending his share.

# Chapter 14
## Trib 2

Monday morning, our second day on Trib 1, was misty and dew-laden. When we arose, the sun was already high and beginning to burn off the haze. Our first activity was a shower down in the stream. It was exhilarating in the clear cold water, and just for a few minutes, our camp resembled a nudist colony. We also laundered a few items of clothing and hung them on a line to dry. Only then did we set about preparing breakfast.

After breakfast, we rounded up items we wanted to take and packed smaller items in the nylon day pack. The Koolaid, a few snacks, the folding shovel, a gold pan, maps, and photos of the valley. Rifles of course, a must, and belts with waist pouches. We debated whether to take our ponchos and decided it was not necessary since the day was shaping up beautifully.

Traveling light, we hiked around the end of the ridge to explore Trib 2. We were able to stay at low level all the way. We knew the valley was wet and boggy so our biggest problem might be to avoid wet areas. With luck, we'd be able to stay on drier land at the base of the ridge.

The ground was firm while we stayed on the narrow strip of land between the treed slope and the wetter valley. As Trib I headed south, we started east around the end of the ridge. The southern tip of the ridge was narrow and defined. The wide expanse of Trib 2 Valley lay ahead of us. Trib 2 lay five hundred yards to the east, across what was obviously wet ground. We stuck to the edge of the valley as we turned northward.

About a mile up the valley, a loop of Trib 2 came close and we were able to make our way to the stream bank. What a difference. The stream had none of the babbling vitality of Trib 1. It was a sad apology for a stream, more like a canal with slow-moving water the color of weak tea.

We theorized it was related to the size of the catchment area. Trib 2 catchment was much smaller in area because it lacked the wide spread ridge as was found to the east of Trib 1. Consequently, the water drained from the hillsides into the swamp and then leached through the bog to form a stream.

There wasn't much of a choice but to follow the course of the stream northward. Fortunately, it was still reasonably dry underfoot on our side and there were no contentious willows. At one point we saw a small beaver dam, with a wide pond retained behind, on the east side of the stream. Actually, the stream was choked with grasses and reeds, but there were open canals visible, indicative of some activity.

It took us about an hour and a half to work our way up to the head of the valley. Meanwhile, the stream became less and less defined. Soon it simply disappeared and we found ourselves in the middle of a bog. We made our way carefully to drier land, at the foot of the hillside, and found a place to rest and have a snack.

It had turned into a perfect Alaska day, with a warm sun, over a spectacularly beautiful pristine valley. There were a few mosquitoes, but many brilliant blue dragonflies flitting around the tops of the marsh reeds. Bees hummed from flower to flower. Trib 2 Valley was not nearly as pretty as Trib 1 Valley, but it was a peaceful place. Given such restful circumstances, we were soon dozing off in noon siesta.

The raucous sound of a Piper Super Cub jerked us from our snooze as it passed directly overhead. Many aircraft, at a distance, flew over during the hike in, but this was the first to come anywhere close. We watched, without moving, as it continued an unwavering course to the east. Only Canada lay that way. We figured it might be heading for Dawson or Whitehorse. It seemed strange however it would be traveling at such low altitudes for such a long flight. Perhaps there was some smaller settlement to the east where the versatile Super Cub could land.

Well, here we were, at the head of a quiet, sunny valley, no way resembling gold country. Of course there may be gold, but where would we start looking? There was no stream to concentrate on and without flowing water we couldn't rig even a basic sluice-box. It had been an enjoyable little hike but we scratched it off our list of prospecting places.

We started to retrace our footsteps south down the valley and reached the beaver pond. Hagen whispered. "I think there must be a beaver." He pointed. "Right there, see the ripples."

We stood quietly to watch. After a few minutes, to our surprise, it was not a beaver but a cow moose which came into sight. We stayed still and quiet and watched while she browsed on the swamp vegetation. Had we had a camera, with a zoom lens, she would have been the star in a beautifully natural picture.

She moved into the bushes again and we set off quietly downstream lost in reverie. Seeing, and not disturbing, wildlife gave us a sense of belonging to this place, this moment, and this planet.

It was four thirty by the time we got back to camp. Everything was as we had left it. We were always worried about bears picking up the scent of our food and cooking and coming to ravage our unattended camp. Not so this time. Our laundry was dry and we packed the clean things away. Then we started a fire and prepared our evening meal. Nothing like a mountain meal, a good cup of coffee, and hours to relax in total quiet splendor.

I had been keeping track of events, in a little red note book, and registering various observations during our hike. I had in mind, to some day, write an article about our exploits in these valleys of Alaska. Frankly it seemed there was almost enough material to warrant turning it into a book instead of just an article.

Hagen peeked up at me and commented, "You'll probably make more money writing a book than we will gold panning."

"How would you know, having never written a book?" I bantered. "Besides, I may be successful at both, then a book would be even more exciting."

As the sun slid slowly into the west, it turned cool and prompted us to keep the fire well stoked. It was dramatically cooler than the previous evening and it reminded us that weather in Central Alaska was fickle and could change very quickly. It was not unusual for some areas to have snow flurries in July. Though it wouldn't stay around very long, it could make things decidedly uncomfortable if we were without shelter. We hoped it would stay nice at least until the remainder of our adventure.

It was about as dark as it would get for this time of the year. The evening was turning into one when we would be more comfortable in sleeping bags than huddling by the fire. Once we were in our sleeping bags we, as usual, found something to chat about. Somehow this evening we got around to the subject of ladies.

We started talking about the gals we knew through dancing and aerobic classes. We began to speculate about meeting Mrs. Right. For sure, it would cramp our present liberal lifestyle. We expressed some emphatic reservations on that score. We knew a lot of women but, sweet as some of them were, running down the list, we couldn't think of one with whom we really had a chance of a long term relationship. There were a few we fancied, but they had others in mind. Obviously, the chemistry just wasn't there and Mrs. Right, just as well, hadn't come on the scene for either of us yet.

I mused, "Maybe we'll be lucky enough to meet them out

here in the wilds, two rugged, outdoors, suntanned beauties. Maybe they'll be bent on hiking and gold prospecting like we are."

Hagen turned over and snorted. "They may turn out to be just as ugly as that moose we saw today, too."

"C'mon, Hagen. Maybe they're camping over in Trib 3 Valley. Maybe we'll meet'em tomorrow."

Hagen huffed. "Hey listen, there are gold diggers and then there are gold diggers, and most of the female prospectors, if there are any, are not likely to be raving beauties. You just stick to that little red note book and let me do my own prospecting."

We lapsed into silence with the certainty there would only be the two of us for breakfast the next morning at Hotel Trib 1.

# Chapter 15
## Glory Hole

It turned out to be a very tough day. After breakfast we packed everything, and could see it would take two trips. Tools were too cumbersome. Tarps and canned goods added to the weight. In the end, it took two trips and most of a day to make the transfer. By evening, we were absolutely worn out. There had been no climbing involved but the terrain was difficult to walk on with heavy and awkward loads.

Once everything was at Glory Hole, we secured a tarp between two trees as a shelter. We made our camp on the eastern side of the stream above the bedrock ledge. Then after all of the sweaty effort, we decided another bath was in order. Again we braved the icy cold water of the stream. It was tough all this hiking and prospecting and having to wash in cold water. We cursed and complained but we enjoyed ourselves.

When we were dressed, we put together a tasty meal of beef stew with extra vegetables and sliced Spam. Dessert was a can of sliced peaches. Trib 1 water made excellent tea and I had many cups. We firmly believe every stream, depending on its origin, imparts its own particular flavor to tea and coffee. Some, were not so good. Trib 1 water rated pretty high on the list.

"Say." I said to Hagen. "When the gold plays out, we can always go into the bottled water business. Imagine super markets stocking gallon jugs of Genuine Trib 1 Water. We could make a fortune!"

Hagen didn't even look at me when he said, "Stick to your red note book."

Our chosen campsite, at the foot of the steep slope, was fenced by trees on the north and east side. We felt a bit uncomfortable about limited visibility. Our site was elevated a bit and we had a view over the crescent shaped claim area and rounded hillside to the west. Downstream of the camp, we could see the fallen tree bridge crossing Trib 1.

It was an attractive location because there were colorful wild flowers around the area. In open areas of the slope, sorted masses of red fireweed swayed in the breeze. We would loose the sunshine earlier in the evening due to the ridge—a real disadvantage. It was almost idyllic: cascading water two hundred yards upstream, babbling of the small stream as it made its way past the campsite, and sound of bees buzzing from flower to flower.

Hagen said, "If we decide to work the claim, we'll have to put up a better shelter."

I agreed, "Right here, where we are is as good a site as any." As we suspected, our campsite lost its sunshine a bit early in the evening then quickly turned cool. After such a strenuous day however, we called it quits and crawled into our sleeping bags quite early.

Despite our fatigue, neither Hagen nor I slept well. We spent the whole night hearing movements in nearby trees. At times, I found myself straining to hear every little rustle. I'm sure most noises were made by small rodents or by leaves or twigs falling from trees, but my mind converted each sound into sneaking bears. I was almost glad when morning finally arrived.

Hagen was, if anything, more paranoid about bears and suggested we stoke up the fire at night as a deterrent. We weren't sure it would make any difference, but there was some psychological advantage to the idea.

Bleary eyed, we had an early breakfast. Lots of coffee brought our courage back and we soon felt a bit better. A slight breeze had taunted us most of the night. No mist in the valley this morning. We decided we would do some serious prospecting, and then, if it still looked promising, we'd measure out a twenty-acre claim area and encompass the Glory Hole.

We marked a grid over a crescent shaped area simply by starting at a point on the northeast corner and then, over the prime claim area, mark five yard by five yard squares using rocks and stakes as markers. Then, we started sampling the gravel at each marker to a depth of two feet. With the pick and a full sized shovel this was an easy task. We sketched the grid on a sheet of paper and recorded our observations for each coordinate.

It was time consuming, but we worked our way over the area recording the depth of the gravel, size, coarse, medium, fine and any obvious false bedrock. A false bedrock was a layer of impermeable clay. We never hit actual bedrock, but believed it was not far. At each site we panned small samples from different levels to see if there was any gold. At most locations we found a little color which further reinforced our positive feelings about the claim. We were convinced, when from one location close by the rock ledge, we found a grain-sized nugget along with fines in the bottom of the pan.

By early evening, just in sampling efforts, we had collected about a quarter ounce of gold. It was marvelous. We had panned only a very small amount of material. If we had sluiced, say a cubic yard of this material, we may have obtained an ounce or so of gold, making this a better than average placer claim.

It took us sometime, some crude triangulation, and much effort to measure out a twenty-acre claim. It covered an area four hundred and fifty yards downstream on the grid, three hundred and seventy-five yards upstream, and two hundred twenty-five yards wide. We drove four good stakes to mark the extremities of our claim and nailed a piece from our map, suitably marked and covered by a Ziplock bag, to the post at the north east corner. We also placed a notice at the Glory Hole showing the extremities of the claim. We then worked to tidy up the rest of the area. We didn't want to leave more traces of our activity than were absolutely necessary. The laws would then be satisfied to the marking of a claim.

Now we had a good idea of what lay underfoot, at least to a depth of about two feet. What we really needed, but didn't have, was a long crowbar with which to probe to bedrock. By the time we finished, it was six PM, and we were hungry,

It was a fine evening. We ate our dinner and surveyed our claim and we were satisfied with a good day's work. We decided to really work on one spot. More gold would go into the vial tomorrow. We had to reward ourselves a little bit for all of the effort. We whooped and hollered and bragged to a silver moon. We had gold fever which needed to be satisfied, so we danced and romped around the campfire.

We banked up the fire, as Hagen suggested, placed some extra logs nearby and headed for our sleeping bags. It seemed a bit risky but at least we might sleep better. It turned out we never used the spare logs and the fire was cold coals the next morning.

## Chapter 16
## A Day of Panning

A light breeze stirred the tops of tall spruce trees and bright sunshine filtered through the undergrowth. It was a wonderfully, beautiful morning—our sixth. We felt smug about our trip thus far. Our small fire, sufficient for boiling water, crackled and glowed as we launched into a day of serious placer mining.

Research the previous day showed we could find gold nearly everywhere we looked. We decided to be scientific and find a more productive spot. The bedrock cliff acted like a sluice-box. Gold may have settled deep in back water clefts of the bedrock cliff. We selected a likely place and started digging. The little V-shaped cleft had an area of five square yards and was filled, just above the present stream level, with moderately fine gravel and sand. With care, we stayed out of the water, shoveled away two feet of the overburden, took material into our pans and washed it in the stream.

We quickly found it uncomfortable work. Trying to keep our feet dry caused us to bend over the water, which was lower than the bank. It would have been more comfortable to wade into the stream to carry out the panning. Though our feet and legs got wet and cold, the lure of gold made us forget the discomfort. We worked our way downward in the cleft, an inch at a time it seemed. There was gold in almost every pan of material.

After an hour or so it dawned on us that we were spending a lot of time washing the contents of our pan down to the point where only the gold was left. We realized we might also be losing fine gold flakes in the process. We started panning until we had a small quantity of heavy sand and gold flakes left in our pans. We saved these fines in a Ziplock bag. We'd have to pack it out or leave it here on the claim, either way, we could spend time later carefully separating the two. It made it difficult to estimate just how much gold we really had but we could tell there was a lot. Once we adopted this sensible practice, we moved more material in the time we had available.

Throughout the day we rested but managed to put in a total of ten hours of digging and panning. At the end of the day we had two Ziplock bags of heavy black material and what looked like an ounce of gold. With aching backs and knees we finally called it quits. Not to attract attention, we filled the flooded hole and cleaned up the site. We spent a little time arranging items we wanted to stash on the site. Scouting around, we found a safe place. Hagen found a recess, almost a small cave, under a four-foot-thick slab of protruding bedrock uphill behind camp. We hadn't noticed it before because it was well concealed from our campsite by a stand of squat spruce trees. It was ideal. It had plenty of room to house tools, tarps, and any left over supplies. We gathered logs to cover the entrance. The contents could remain there until next summer.

Dusk arrived early. A mass of clouds had gathered in the west. The weather was about to change for the worse. Rain tonight would make our hike to Trib 3 the next day more difficult. We sat close to the fire and enjoyed the warmth creeping back into our legs. The coolness of the evening demanded jackets but we were glad to relax after such a back breaking day. We boiled a gallon of water and set it aside for the next day's journey. We were under no illusions, it would be a tough day, starting with a long climb out of the valley and up to the main trail.

Morning seemed to come all too soon. I was awakened by rattling cookware. Hagen was up and about and obviously hungry. I stretched and mumbled, "I just closed my eyes. Where did the night go?"

There was a high streaky overcast, and no sun with high winds aloft and a little cool. The weather had changed, and not in our favor. At least it wasn't raining—yet. Judging by the feeling in the air, it wasn't going to start soon. It could turn out to be a good day for strenuous activity, better to be a little cool than too sunny and hot.

When breakfast was finished, we wrapped the tools and surplus canned rations in tarps and carried them fifty yards up hill to the small cave. Packing them as far into the cavity as possible, we wedged logs in front to provide further protection. It was a good, dry hiding place which, with a little work in the future, might be turned into spacious storage for more than just a few items. Once the winter arrived, and it arrived early in these parts, everything would be blanketed by snow and there would be no evidence of us having been here.

Within fifteen minutes we had our backpacks ready—we decided to take the two Ziplock bags of heavy fines with us—and we took a last look around the area to make sure nothing was forgotten. Then, with rifles in hand, we set off.

At first we made our way up stream through our claim. We stayed as close to the course of the stream as the thickets of willow and alder allowed. A half mile of gentle incline gave us chance to warm up to the idea of hiking with a full load again. The terrain steepened. We weren't ready. At the same time the stream we were following split in two. We followed the right branch which led us toward the main ridge.

As we ascended, the land was better drained. Foliage changed dramatically. Arctic spruce, silver birch, and aspens yielded to large Sitka spruce and there was considerably less tangled undergrowth to negotiate. To ease the grade we made long zigzags across the slope. We wanted to see if this was a more feasible way to access Trib 1, so we paused occasionally to take note of position and prominent features. So far, it was good. We didn't see any features which would prohibit the use of some kind of all-terrain vehicle.

Straining under the considerable weight of our backpacks and rifles, we climbed steadily upward. The occasional rests were welcome. We'd catch our breath then move on. The surrounding countryside lay before us, as did the bulk of the western ridge. Despite the high cloud layer, we could see a long way even to the clearly defined peaks of the Mentasta.

The hillside turned out to be a series of terraces which at times, gave us a break from hard climbing and enabled us to meander our way with ease. As we gained altitude, the trees thinned and a swath of six feet high willows impeded our travel. It was tough going for a quarter mile. Eventually, with the steadily increasing altitude, the willows ended and we broke out into the lower growth heather and berry bushes of dry arctic tundra.

Here, on the exposed hillside, the valley breeze changed into a minor gale. By the time we reached the main trail on the crest of the ridge, we were straining against the wind. We decided to take shelter on the lea side of the ridge and break for lunch. A hundred yards along the trail, we found a suitable place for a camp in some low growing spruce.

Thirty minutes later with our snack finished, we lay back to rest surrounded by low growing arctic heather and blueberry bushes. We dozed to the soothing sound of wind whistling through bush tops.

An hour later, feeling refreshed, we moved on. First however, we needed to stash the package of food and water for our return trip. These supplies—and the heavy fines—had been split between our two back packs. We placed them into one plastic trash bag, knotted the top securely, and placed the bag out of sight at the base of a gnarled spruce tree near the trail. Now we had two caches on the trail and wouldn't have to worry about food or water on the hike out. We felt pretty smug.

Setting off with lightened backpacks, we bounded in a northeasterly direction along a nearly level trail. It was like a major highway compared to the terrain we had been on for the last few days.

After a mile or so the trail steepened and brought us to a more sensible pace. We were ascending gradually toward the highest point, where we were a few days earlier. Then, it had been misty and we had only limited views. Today was considerably better and when we reached the summit it seemed as if we were on the top of the world.

For many miles, there was nothing higher than us, though some ridges were almost equal. Our vantage point provided views in all directions, to the Wrangells, Alaska range, and to Mt. Fairplay. There was also an excellent view of the trail continuing to the northeast, the direction we'd head for a couple of miles. We could just about discern the point where we'd leave the main trail and venture down into Trib 3 Valley.

It was disturbing to see thickening clouds scudding overhead. They were still high but it did not look well for the next few days of hiking.

We took another break on the lea side of a rocky outcrop. There we could enjoy a view to the north. We tried to stay out of the chilling wind. Consulting our map, we noted the Trib 1, Trib 2 ridge was due south of our position. Trib 3 ridge was about one mile away to the east.

We realized we were sitting in the middle of a particularly good patch of very juicy blueberries. Only after we had enjoyed dessert and collected a few berries in a Ziplock bag, did we hoist our backpacks and rifles and start our hike downhill.

The strong wind on the exposed ridge unbalanced us at times, but the fresh air was exhilarating and it helped us on our way since it was pushing us. It took us only thirty minutes to reach the point where we would have to forsake the beaten track and take to virgin territory again.

We paused to compare our most detailed map and the lay of the land. It was immediately apparent that it would not be an easy hike down into Trib 3. At first the land fell gradually into a profusion of willows and other shrubs difficult to negotiate. Where the larger trees started, the land fell off steeply. From our vantage point it seemed best to traverse the slope, for a short distance in a southeasterly direction, turn south through what looked like a narrower band of willows, and then cross a steep slope in a southwesterly direction. It was a longer zigzag course but it would be easier.

Strange, but whenever we surveyed such an area, it always gave the impression of being prime bear habitat. I guess we had been extremely lucky so far. Well, we couldn't stop now so we checked our rifles, cinched up our backpacks, and set off bravely down our planned route.

Five hundred yards downhill, we entered thickets of six foot high willows and had a difficult time negotiating our way. Here and there we picked up a trail, probably made by caribou, and found a little relief, but for the most part, it was a struggle through tangled foliage. Pointing ourselves downhill we blindly fought our way yard-by-yard until we reached the tree line. The willows ended abruptly, walking became easier, and our range of visibility improved.

As we descended, the trees became larger. We were just talking about it when we noticed great gouges in the bark of a tree directly in our path. Our fears were instantly confirmed, this was prime bear country. The bear with that size of claws was a big one. It was a stretch for us to reach up to the deep scars where the bear had manicured its nails by raking the tree trunk. Big shards of bark lay on the ground around the base of the tree. Nervously, no, very nervously, we looked around before picking up bits of bark to examine them. They didn't look fresh but they weren't old either. A few days old at most we thought.

A bear could travel a long way in a few days. We had no way of knowing if it had left the area or if it was still nearby. Judging by the span of the claw marks, about eight inches, we decided it must be a grizzly. We didn't consider ourselves experts. Could've been a large black bear. With our senses jolted, and on alert, we continued noisily downhill with rifles at the ready, hoping the bear, if it was still in the area, would decide to relocate.

The slope steepened noticeably and we had to skirt embankments and rocky outcrops. Our backpacks and rifles were cumbersome, and hanging on to branches or tree roots to keep our balance was extremely difficult. Progress was slow. An injury here could be disastrous. We were further from the highway than we had ever hiked before and the logistics of hiking for help and arranging a medivac out would have been a daunting prospect. We were extremely careful.

Soon, through the trees, we saw the outline of another ridge and we realized we were descending into a narrow ravine, an offshoot of the main valley. There was the gentle burbling sound of running water below. Fighting our way through thick undergrowth we discovered a small stream of clear water tumbling downhill between mossy banks. We followed the stream downward, torturous though the terrain was, until the land leveled-out near the valley floor. We left larger trees behind and were surrounded by smaller aspens, silver birch, and a tangle of tall alders.

The head of Trib 3 Valley was formed by a deep V in the hills. The hillsides were much steeper than those around Trib 1 or Trib 2 and the valley floor was narrow with little room for swampy ground. The little stream we followed eventually joined the larger main stream.

When we reached its junction, we decided to take a rest. It had taken us almost two and a half hours to make our way down from the ridge and most of it had been tough going. Exhausted, we were glad to take off our backpacks, build a small fire, boil water for coffee, and make dehyd stew.

Kicked-back and resting a short while later, letting our lunch settle down, we realized the cloud layer overhead looked more threatening. There was a chill, a feeling in the air, a prelude to rain. It might pass us by but we felt certain we were going to get wet before much longer.

While we rested, about an hour, we decided to explore a little upstream and then go downstream to spend the night. To make it easier we took only the nylon day pack with one gold pan, the aluminum shovel and our Gortex jackets. Of course we were not about to move very far at all without our rifles.

The main stream was about the same size as Trib 1 at the Glory Hole. Right or wrong, the best way to follow the stream was via a defined animal trail. There were no fresh prints but it didn't guarantee safety, so we made lots of noise and were constantly on the alert.

The clear water stream curved eastward, babbling in a stony bed. After a quarter mile or so, the valley steepened and the stream bed became even more rocky. There were times when we couldn't see the stream at all because it was coursing under a jumble of fragmented rocks. It didn't take a geologist to see that most of the rocks were fragmented and unworn, not at all like the rounded boulders and pebbles of Trib 1. Deposition of loose materials here was due to landslides and the freezing and thawing action of the seasons rather than hydraulic action.

We dug a few shovels of sand and gravel at convenient places and panned to see what we could find. We detected no gold particles at all and finally, as the stream shrank in size we gave up and retraced our steps to the backpacks. We delved into our packs for our down vests. It had turned cooler. Though we had not been long on that exploratory hike, we craved a cup of hot tea. So we started another small fire on the ashes of the first and boiled more water. The hot tea and the warmth from the fire was comforting.

Looking around we had to agree the valley was not very inspiring. It was deep, overgrown and dank compared with

Trib 1. Maybe downstream where the valley widened it would be a little better.

After a thirty minute break, we moved on. This time we took all of our gear. We'd have to find a place downstream to camp for the night. The animal trail—naturally—followed the stream most of the way and rifles at the ready we rarely left it. The valley widened perceptibly, the trees thinned and became a little more pleasant.

There were clumps of silver birch and aspens and open areas filled with tufted marsh grass. Batches of larger trees, mainly conifers, were separated by sheer rocky cliffs.

Continuing south, the trail became well defined. The trail had been made by some kind of tracked-vehicle. A vehicle very different from those monsters which had cut the main trail on the ridge. It was obviously a small, lighter vehicle, the tracks barely six feet wide, and the type favored by hunters and prospectors. We were at a turnaround point and the vehicle had not traveled any further north. The tracks did not appear to be recent and possibly were even made last summer. Clearly, we were not the first to enter this valley. These tracks must surely be the same ones we had spotted from the air during our reconnoitering.

Though the tracks were old, we went carefully. We didn't want to go blundering unannounced into someone's camp. That could be dangerous here in gold country. After about a quarter mile we came upon what had been a well-established camp site. There was a hearth made of large river stones and firewood stacked nearby, as if awaiting someone's return. There was no sign of any other equipment and the vehicle tracks from the south were eroded and washed out. No one had been to the camp this summer.

About fifty yards downstream of the camp there was sign of placer mining activity. Large excavations in the western bank of the stream hadn't been cleaned up and two trees had been dropped across the stream as a foot bridge. Strange, there was no settling pond for the tailings. Even in the remote areas, miners sluicing large amounts of material, were expected to comply with basic rules to prevent fouling streams. Also, there was no sign of equipment, though we supposed it could be stashed nearby. What about claim markers? We hadn't noticed any on our way downstream. Had the miners simply not bothered?

Continuing southward, more watchful now, we eventually found a weathered marker nailed to a tree trunk. It was barely readable but we could see the claim area consisted of six claims covering some distance up and downstream. We had obviously missed seeing some more northerly markers.

We had walked maybe a quarter mile further south when Hagen, in the lead, suddenly froze and let out one of his favorite expletives. About fifty yards downstream was a big black bear. We stood still, surprised and a little scared. All this hiking and now we see our first bear. Well, we had walked on enough of their trails it was probably long overdue.

It seemed we were the only ones excited or scared. The bear seemed to ignore us and even managed to look uninterested. It stood on the opposite bank of the stream and sniffed the air with a nose flecked with ginger colored hair. The rest of its coat was a beautiful shiny black and not a bit bedraggled as one might imagine. I wouldn't claim to be an expert on bears but this one was a fine specimen.

We stood perfectly still for a couple of minutes until the bear finally moved. To our utter amazement it crossed over to our side of the stream and turned in our direction. I suppose it would have looked funny on film but this was Hagen and I, and we were in the path of the oncoming bear.

In a couple of seconds of blind panic, we almost ran into each other before we realized that, unless we wanted to defend ourselves, there was only one way to go, backward. We retreated slowly, while keeping an eye on the bear, which fortunately was still moving nonchalantly. I would never have believed it. I guess I always thought a bear would either run at us or would run away. This one was out for a stroll in the country and we brave frontiersmen, but with rifles at the ready, were in a none too organized retreat.

Retreat was the sensible thing to do. We didn't want to antagonize the bear and we didn't want to shoot it unless we had to. We were, after all, invading his territory and using his highway which unfortunately was only one lane wide, and worse still, didn't seem to have many safe-passing places.

The bear moved so slowly we were able to gain a little and finally came upon a place where we could leave the trail and splash across the stream. Now, with our backs against the steep hillside, we had to make our stand. We watched with trepidation as the bear came into sight, paused a moment, sniffed our trail in our direction, and then moved out of sight up the trail toward the mining camp. Talk about treating us with utter contempt.

Had there been anyone else within a hundred yards I think they would have heard us both take our first gulp of air. I felt sure my knees had been knocking together for the last few minutes.

Hagen, just as scared as I, commented, "Good thing it was only a black bear."

"Hagen," I said, "May I remind you people over the years have had their rear-end anatomy rearranged by 'just a black.' It is said they can be more dangerous than grizzlies." Anyway, it reassured us on one point, Hagen had no problem seeing real bears. This was not just another tree stump or rock looking like a bear.

Well the excitement was over. After waiting a respectable time, we cautiously crossed back over the stream, and with frequent backward glances, continued our exploration of the valley. The valley was obviously spoken for, but we wanted to satisfy our curiosity and find a place to camp for the night. It would be good to be settled early because the weather was still deteriorating.

We stuck to the ATV tracks until we reached a spectacular rock slide which came right to the waters edge. The vehicle tracks forded the stream but we couldn't do the same without getting wet. We scrambled across the base of the slide to larger rocks forming stepping stones across the stream.

Back on the trail, we continued south for half a mile. There was no further evidence of mining, so we concluded the owners of the claim, like ourselves with Trib 1, had sampled the stream and finally chosen one spot to work. Judging by the amount of material they had moved, they must have been finding gold and would probably return to work it again later.

We had seen the faint trail leading to the valley during our scouting flights. After consulting our maps, we concluded it led from the eastern side of the ridges, either from Sixty Mile on the Taylor Highway to the north or Northway Junction to the south. Both were nearly fifty miles away. It must have been quite a journey, even with a tracked vehicle.

Our map showed the valley widened out soon and became flat land draining down to the Ladue River. There wasn't much point in continuing further south so we decided to camp right where we were for the night. It seemed as good a place as any. Thanks to the days strenuous activity we were both feeling well ragged out anyway.

Not wanting to leave obvious signs of our visit to the valley, we selected a level place twenty yards or so from the trail. As usual we set up our two ponchos close together so that we could talk. There was a plentiful supply of firewood and in less than fifteen minutes we were preparing our evening meal. A large kettle of hot tea brewed from well boiled Trib 3 water with its distinctively different flavor.

Dark clouds shortened the evening and an occasional spot of rain carried in on the breeze as we took care of cleaning away our supper. It was cool but we both had a good sponge down by the stream. Fortunately, the coolness of the evening was suppressing the mosquitoes and allowed us to make the most of the campfire. Thinking ahead, we boiled a gallon of Trib 3 water for use during our next day.

Retiring to the poncho, around ten o'clock, it crossed my mind the bear might be watching quietly from nearby trees. I told myself, "This was a friendly bear." Of course, the rifle lay close beside me and acted as a security blanket. I slept very well.

# Chapter 17
## One Mean Wet Saturday

When we woke at seven o'clock, rain was pattering softly on the ponchos. We debated whether to get up at all. Finally, we struggled getting dressed in the confines of our respective ponchos. We slipped into Gortex raingear and crawled out of the shelter. It was a mighty, soggy world. Vegetation was soaked and dripping. It looked like there was plenty more rain still to come. The clouds were ominously dark and low and shrouded the tops of the surrounding ridges.

Fortunately, we had placed a small store of firewood under the shelter of the ponchos and soon had a fire going for breakfast. It was miserable, there was no sign of any improvement, so there was no point in wasting time. After breakfast, we quickly broke camp and started our return journey.

The night before we had discussed our return and decided to deviate from the original plan. Instead of trying to retrace our steps of the day before, we'd climb to the spine of the ridge separating Trib 3 and Trib 2. As we observed the day before, the hillsides were too steep to consider climbing. However, there was one place, upstream of the rock slide, where the hillside slope was more gradual. We figured we could make it. It was unfortunate that everything was so dripping wet. The slope was the shortest way up to the bald spine of the ridge where we would be clear of wet vegetation. We had to face the fact that every way would be wet.

Following this plan, we hiked north to the rock slide. After studying the hillside, to decide the best route, we struck off into the trees and undergrowth. It was difficult as soon as we left the trail. Low level shrubbery grew in great profusion and was hard to find a way through. The fact that everything was so dripping wet just made it worse. The grade was severe and small embankments became major, slippery, obstacles to upward progression. We gave each other help and used roots and branches to haul ourselves over the worst slopes. Every blade of grass, tree branch, and leaf, flipped or dripped rain water, and in spite of raingear, we were soon wet all over. It was an unkind, steady drizzle.

The grade probably averaged forty percent with steeper sections giving us a lot of trouble. Several times we came up against rock cliffs which couldn't be scaled. We'd traverse across the slope to skirt around them. Tumbled piles of fractured rocks, overgrown with shrubbery, were particularly dangerous to cross.

Under this stress, a tenacious Hagen was emerging. Peering from under a dripping rain hat brim, Hagen's blue eyes gleamed. "Now, here is a challenge, Doug." His jaw was set and I could tell he was visualizing himself as an early pioneer who had to cope with miserable adversity, if they were to conquer the frontier. Trial by fire and all that.

I was also interested in conquering this dratted hillside. Perhaps in a different way, because once we were committed, I just wanted to be done. We were busy climbing and struggling through the shrubbery. We weren't looking around and could walk into a bear without seeing it. The rifle, which was such a comfort at times, was now just one more cumbersome item.

Thankfully, often we rested and regained our breath. Slowly we gained ground. To our advantage, the foliage changed and larger trees began to dominate. This smothered the smaller growth and the going was much easier. We took a lengthy rest under partial shelter of larger trees and broke out trail-mix and Koolaid. We had climbed six or seven hundred feet from the valley floor and had another five hundred to go to reach the bald ridge.

Moving on twenty minutes later, we found the slope lessening just a little and we made good headway for a while. Then, as larger trees gave way, dwarf willows took over. Just when we lost what little shelter the trees provided, we encountered wind driven rain. There was no choice, we just pushed our way upward, taking advantage of small clearings or any pathway. It was just awful. Everything was soaking wet, dripping on us—rain driving into chinks in our clothing. The brush constantly snagged us, our backpacks, and our rifles.

It took us an hour to pass through this atrocious tangle of willows and brush. What a relief when we reached the upper slopes where only bearberries, blueberry bushes, and heather grew. Resting for a few minutes, we hunkered down against the slanting rain. Looking back down, we could hardly believe we had fought our way up that terrible slope under such conditions. It was crazy, but we couldn't just stay there in the valley waiting for it to stop raining. Who knows? Maybe it would rain for a week.

We hardly had a dry spot on our bodies, and the wind was chilly up here on the ridge. We needed to get moving and find ourselves a place to rig a shelter. Number one priority now was to get dry and warm. People died of hypothermia under these conditions and we weren't going to be one. We agreed we were pretty miserable, but we felt okay to continue along the spine to the main trail. It was about a mile in open country. If we were steady on, it shouldn't take too long.

The walking now, was much better than we thought it would be and the mild exertion kept us warm. With rain coming from behind us the Gortex did a good job and at least we didn't feel as if we were getting any wetter.

Thank goodness. The main trail at last. The worst was over. To make our way out, we only had to stick to the trail. If it continued to rain, at least we wouldn't have to fight our way through any more brush.

Right now we had other priorities. We needed to rig a shelter and get into dry clothes. Working quickly we rigged both of our ponchos between two stunted trees. They formed a wide lean-to; a shelter against the wind and rain. Then, still in our wet clothes, we scavenged around for firewood. Long ago we learned how to get a fire going in wet conditions. We used little dry birch bark kindling stashed in our backpacks.

The fire, which we made near the lean-to entrance, was slow and smoky and we coaxed it along. As it became established, we ducked under the shelter, struggled out of our soaked clothing and pulled on our spare dry set. We shook the water off the Gortex and hung it to drip-dry under our small shelter.

The fire was not large but we kept it stoked and its meager heat radiated under the poncho shelter. We boiled water and made hot, extra sweet tea and stirred packets of cream of chicken soup. We had only snack food left; a few crackers, a few granola bars, and some trail mix. We spread it all out and slowly munched. Huddled under the shelter, away from the wind and rain, we waited for the warmth to sneak slowly back into our bodies.

It took two hours for us to recover. When the rain slackened, we decided to move along the trail to our cache of supplies and then make camp for the night. A hard, torturous, seven hours was enough for one day.

Before leaving the tiny shelter, we put on our damp Gortex, making sure everything was well closed to prevent water from sneaking into our dry clothes. Only when we were zipped up, and water tight as was possible, did we take down the shelter. To save time we shook the water off the ponchos, folded them and secured them over the top of our backpacks

It was two miles to the food cache and we covered the

distance leisurely. The plastic trash bag containing our supplies was undisturbed. We had a supply of food and water aplenty for the night and enough to reach our next cache. Under the adverse conditions we were glad of our advance planning.

The rain and wind abated as we began to set up our camp, but there was no sign of a real break in the general lousy weather. It gave us just enough relief to have our evening meal and get ourselves and camp arranged for the night then it started again. Not quite so decided as before, but rain nevertheless. Thankful for small mercies, we rolled into our sleeping bags, under our ponchos, now with edges staked down, and fell into a deep sleep of utter exhaustion. This had to go on record as having been a wretched hiking day.

Before poets romanticize about sunrises, they need to spend a rainy night, under a poncho, on a mountain ridge in Alaska. Our 'dawns early light' began at seven o'clock—it was cold, grey, damp, and miserable. The mist was moisture laden and was just like rain. Visibility was down to a hundred yards. There was no indication of a pending change.

First, we lit a fire and boiled water for breakfast. Then, we stoked up the fire as much as possible and attempted to dry out things a little while we had breakfast. It was a bit futile but the roaring fire was good for morale. It was not the kind of morning which encouraged loafing. So, we packed everything carefully, scooped water from a puddle, and doused the fire. We set off to the west.

There was no view to admire and nothing to delay us in any way, so we just plodded along solidly for miles in silence. With the thick mist close, it was difficult to tell at times where we were, or if we were going up or downhill.

At eleven o'clock, we took a break and made a small fire sufficiently to boil water for tea. It was important to consume something hot to ward off the damp chill. Despite the weather, or maybe because of it, we were making pretty good time. At this rate, we should be able to reach our second cache at Poplar Hollow by mid-afternoon. From that point, to trails end, was still more than we could manage in a day, so we hiked for a few more hours into the evening. Perhaps we could make it to the high point and the attractive camping spot where we had rested on the way in. That would leave us with one short day of hiking to the Chevy.

There was marginal improvement in visibility but the cloud layer overhead remained. It seemed as if it would start to rain again any minute. No sooner had we commented about this when it started to sprinkle. We hurried to round up our gear and hit the trail before it started in earnest.

It rained, but at least the general visibility was slightly improved. We carefully arranged our raingear to keep out the water and slugged onward. Soon we recognized a long undulating stretch which preceded the long climb to the first high point we would have to pass over. On the other side was the steep shale slide which had been so difficult on the way in. It was a long tiring climb, an hour and a half to the hill top, but at least there was no brush to give problems.

We held our rest to the minimum and then crossed the hill's flat top to the shale slide. This slope would be a definite obstacle to any kind of vehicle we might consider using later. We would have to find a more gentle route. If conditions had been better, we would have scouted for a better way, but right now we didn't want to go plunging into any more wet brush.

Getting down the slide was difficult and painful. It was wet and slippery and we both lost our footing several times, bumping our rear ends, elbows, and rifles. We picked up a few more bruises too. We discovered them later. We finally made it down to more level ground, sore and muddy.

By way of reward, it was a couple of easy miles to our supply cache at Poplar Hollow. Eureka! Water and food. We divided it between our back packs. From here, it was a gentle uphill slope with no steep sections all the way to the crest of Mt Son. The rain abated and by the time we reached the summit, it had nearly stopped and the sky was considerably lighter.

We started to set up camp between the little grove of gnarled spruce trees and the rocky outcrop. While these preparations were going on, the sky brightened, and at last we removed our raingear. We rigged a line and hung wet clothes. Except socks, we had no dried spares left. Ours was a tenuous situation. It was important to dry clothes, if possible.

Scouting around spruce trees, we gathered enough firewood to keep us going for the evening, plus a little for the morning. We then prepared a well-deserved dinner. An ideal camp set up, trees and rocks sheltered the site, ponchos were set equidistantly from the fire place, our cooking pots hung from the quadrapod and our wet clothes swayed on the line in a stiff breeze. This, our last night on the trail, might not turn out to be too bad after all.

We put together an excellent meal of dehyd beef stew, dehyd vegetables, and a big canteen of hot sweet tea. I surprised Hagen by digging into my back pack and producing two small cans of fruit cocktail for dessert.

"Hah. I was right." He said. "You're a sneaky cuss. What else you got in there you haven't told me about? Never mind, where's mine."

I laughed at him as we savored the fruit, cube by little cube, "Pineapple from Hawaii, peaches from Georgia, apples from Washington" I counted them out.

"Don't know what this is, probably a pear. Ha, here's a grape from California." Hagen swallowed the last. "And to think we are in the mountain tops of Alaska enjoying such a thing." We both laughed.

There was a damp chill in the evening air so we stoked the fire and huddled close by to keep warm. Late in the evening the breeze pushed away some clouds and we saw a few clear patches of sky. This was more like it. What a difference from the miserable rain we endured for the last two days.

We sat late into the evening, reminiscing about the hike,

the claims, the bear, and the challenges. What will we do next? Hagen and I agreed we should eventually buy an all-terrain vehicle to ferry in sufficient equipment for working The Glory Hole. We'd shop around. If we bought something soon, we could spend time during the winter getting ready for another foray. I didn't attach too much to dreaming, but I could see Hagen's brain spinning into high gear. The pioneer gleam came into his eyes. I knew him well, and something was cooking.

At eleven we called it a day. It had been a tough one and we were both dog-tired. Minimal ablutions, but what the hell, we only had to survive one more day.

It was lovely and quiet. No sound of running water, no rain and no wind—not even a buzzing mosquito. Just the absolute peace and quiet of an unnamed mountain in wilderness Alaska.

## Chapter 18
### Last Day of Hiking

Our last morning. Sunday. I drank the last of my tea and told Hagen, "We'd better start packing for the hike out." The weather was still fine and sunny, but a little cool. It would be a pleasant day for hiking. The trail ahead was lined with just a hint of fall colors. There was some showing on higher ridges too. Bearberry leaves were crimson and blueberry leaves were orange. For a few weeks the slopes would be adorned with beautiful rustic fall colors. The snow would come to this region, and everything would be blanketed until next spring. This very mountain top would be a very bleak and inhospitable place from October until May.

By eight o'clock we were on our way. Backpacks were about as light as they would be and we were unhindered by raingear. First, there was the steep downgrade on the western flank to be negotiated, but it was drier underfoot now and we managed to make it all the way down without incident. We fairly romped along the saddle toward the next peak about three miles away, and bit into the upgrade in grand style. The slope was gentle, but of course, it took the wind out of our sails and soon slowed us to a more sensible pace.

At the top of the three thousand eight hundred and six feet high unnamed peak, the last high point on the trail, we took a half hour break. The clear day rewarded us with a marvelous view in all directions. It was a great change from the confines of the valleys and poor visibility of the last few days.

On our way in we viewed the ridge stretched out for miles. Now it was all behind us and in the distance we could make out the furthest high dome we had passed over. It was no longer just a reference mark on our maps. We had been there, and a little further too. It all looked so far away. Could we possibly have walked that far?

Starting downhill again, we caught sight of the Taylor Highway in the distance. Our first real glimpse of civilization, if you could call a remote gravel road civilization.

It was three and half miles, and a descent of two thousand feet, down to the drainage by the Chevy Blazer. Downhill did not mean it would be all easy. There were steep sections with washouts and eroded areas to carefully negotiate as the trail zigzagged to the lower level.

As we descended from tundra conditions, we could see over tree tops of the lower elevations. There was a great variety of growth noticeably different between the hillsides with a southerly, as opposed to a northerly exposure. It was a land of such contrasts, and some surprises too, such as the grove of poplars we had discovered on our way in and the magnificent conifers down on Trib 1.

With a couple of brief rests on the way down, we soon approached the bottom of the hill. It occurred to us that the swamp might be primed with fresh rain water but we weren't prepared for what we found. The swamp was now more like a pond with a stream running through it. The water was at least six inches deeper than it was when we entered. We already knew there was no practical way around it. It stretched north and south. There was nothing for us to do except to wade through. We had managed to keep dry—to this point. Oh well, the Chevy was only a few hundred yards away and we had dry clothes there.

We set off into the water without even rolling up our pants. We were soon over our knees in water and muck. At one point, straying off the trail, we got stuck in the mud. We held onto each other for support.

"Now would be a fine time to meet another bear." Hagen commented as he hung onto me.

That did it, we started laughing and all but fell over in the swamp. Regaining our composure, we eventually extracted ourselves from the sludge. Uncomfortably, we sloshed the last two hundred yards to where the Chevy was hidden. It was undisturbed.

We set our backpacks down one last time and stretched our tired bodies. We had done it. Over a ten-day period we had walked at least a hundred miles. Granted, some of those miles had not been difficult, but others had been downright miserable considering bad weather and all. Anyway, we congratulated each other on a job well done. We had walked in, located a Glory Hole, staked our claim, explored, and survived the walk out. It was quite an achievement, at least for two middle-aged wanna-be adventurers.

Without changing our wet clothes we removed the camouflage and tarps from the Chevy. We recovered our possessions, keys, and wallets from their hiding place. Just a bit of a groaning start, a moment of anxiety, it fired up and ran smoothly. I coaxed the Chevy out onto the trail. Hagen walked ahead to guide me around any obstacles.

Once we were in the open we stopped and changed from dirty wet hiking clothes into clean gear. Well, we were unshaven and still a bit grubby, but that didn't matter much out here.

Next, we unloaded our rifles and cleaned them as best we could. We hadn't fired a single shot in the time we had been out. Maybe it was just as well. With our gear packed in the rear of the vehicle, we were ready to go. Hagen walked ahead and made sure there was no traffic coming either way as I pressed the vehicle over the roadside berm. Again we kicked loose gravel over the tire tracks to conceal the entrance to the trail.

After the long hike and then sitting around on whatever was available, the comfortable interior of the Chevy felt a little strange at first. We were warmed by the heater and began to feel very weary. By the time we reached the Tanana River Bridge, we realized it would be foolish to drive back to Anchorage this evening. Space available, we'd stop at Tok Motel for a shave, shower, a good meal, and a night's rest.

We were in luck. There had been a cancellation and a room was available. We had a severe case of sticker shock at the price. They condescended to a discount when we gave them an argument. The price was still fifty four dollars, but after all, this was the world infamous Al-Can Highway.

"None of this "me first" stuff, Doug." Hagen shoved me aside when I spoke to shower first.

I shoved back, Oh, so it's you first? I don't think so."

"Then let's flip for first in the shower." The silver dime went high into the air. "There." Hagen said smugly, "I win. Where's the bathroom?"

I thought he was going to spend all night in there. When it was my turn, I understood why. The hot water felt so good I didn't want to leave either.

Feeling much refreshed and certainly looking and smelling much cleaner, we strolled along to the hotel restaurant. During dinner we were surrounded by out of state visitors. It was the tourist season and Tok was on the major migration route. The Al-Can was gradually being improved and more people were making the trip. It was said there would be a hundred thousand tourists crossing into Alaska at Beaver Creek. Never mind the inflated prices; we had T-bone steak and baked potato followed by apple pie and ice-cream. Then we relaxed over an after dinner drink and soaked up touristy conversation around us.

Strangely, neither of us slept soundly that night. We ate too much at too late an hour. And, the beds felt strange after sleeping on the ground for so long. We did however grab a couple of extra hours in the morning and consequently didn't make it for breakfast until nine thirty.

At the gas station along the road, we topped off the Chevy tanks with enough gas to get us through to Gulkana where the price was a bit lower. Then, we went to the State Troopers office to let them know we were back safely.

By chance it was the same sergeant on duty and he laughingly said, "You guys are crazy to stay out in the wilds in such foul weather."

"Well," said Hagen, We'll take you the next time and see if you can keep up with us, foul weather or not."

"Get lost." He said grinning. "Thanks for checking in. Saved me a big search party." Little did he know just how much we had been through.

The drive to Wasilla was uneventful, but the view of the mountain scenery was marvelous. We made only two stops; the first for gas in Gulkana and the second for coffee and a snack at Eureka Lodge. It was after seven in the evening when the V8 sounded its usual klink, klink, klink as I switched off the ignition inthe driveway at Hagen's house.

Our prospecting trip had been a test of endurance. It had been a resounding success. It was only a prelude however. We could work and plan during winter months to make future trips more successful, and hopefully, more profitable as well. We had something tangible to show for our trip too, a couple of Ziplock bags of heavy sand laced with genuine Trib 1 gold.

## Chapter 19
### Hagen's Winter Trip

It was cold. Terribly cold. Every tree limb, every branch, twig, spruce needle, and shrub, sported a thick coating of hoar frost. The land glittered in the thin midday sunlight. Temperatures hovered at -25°. The ground was covered with a blanket of pristine snow, drifted high in sheltered places.

Hagen should have thought about the beauty around him but he was in a terrible predicament. He, his snowmachine, and sled, were tangled in a dangerous situation. It was too cold for any delay.

So far, it had been a lousy trip. Hagen's feet were wet, and cold. His back was wrenched with painful spasms. "Maybe I've bitten off more than I can chew." He debated. His planning, preparations, and practice for this winter trip, didn't prepare him for such a series of problems. Now this. Who would have thought that where everything seemed to be so solidly frozen, water would be a problem.

"Hagen, I said, "It is impractical to buy two snowmachines right now. Besides, I can't get off work for more than a day now, then again next summer." We were drinking coffee in my living room trying to think through a winter trip to Ladue Valley.

"I've already told you I will fly over your route to see if

you're okay. We'll have radio phones too. I just can't be gone for six or seven days."

Silence filled the room and I went to the window and watched the falling snow. "OK, Doug, let's do it that way. I'll go alone with you checking on me."

"Okay. I'll help you gear up. I really wish I was going with you."

Hagen had never traveled during winter to the Ladue Valley and Trib 1. But, under right conditions, he figured it was possible to make a quick trip across forty miles of frozen, and snow covered terrain by snow machine.

His sled carried camping gear and immediate supplies. He carried basic gold mining equipment too. There was an aluminum sluice box, some precut, and drilled framing lumber, planks of wood, a big crowbar, and a long handled shovel. Galvanized buckets, a large galvanized tub, two large tarps and a miscellany of ropes, rounded out the list. Serious placer mining the next summer was his goal.

Being alone in the wilderness didn't bother Hagen in the least. He preferred to be alone most of the time. He had a tough, pioneering spirit and often said he was born a couple hundred years too late. Well, never mind. If it had been done before in the Ladue area, it was a challenge for him and in his own mind, he was pioneering and pushing his personal limits.

His big mistake, he admitted later, was to go alone. Two machines, and a partner would have made all the difference.

It appealed to his enormous sense of adventure to be alone in the wilds of Alaska and the dead of winter and dangerously cold, only added spice to the adventure. He told me later, "I was very lucky." Winter temperatures could get down to -60° in the Ladue Valley.

Hagen drove his jeep with trailer from Wasilla. He camped for one night near the Taylor Highway. There he readied himself and his machine. "That was the easy part." He said. "I slept comfortably in my dome tent and I used the Jeep to warm up during the day. Once I set out on the snow machine, those luxuries were gone. Then my troubles came with a vengeance."

To make his way into the Ladue Valley, he had to pass through four miles of forested land. It was wilderness with no trail to follow. The forest turned out to be a veritable tangle of trees, and between the trees the snow was soft and much deeper than he had anticipated. "I hadn't traveled two hundred yards until my machine bogged down in the deep snow. Hagen said. "I shoveled and packed a trail to get going."

Every hundred yards, or so, he ran into trouble. Sometimes it was because the snow was too soft and deep and sometimes he simply found his route blocked by trees bowed down by the weight of snow.

Each time the rig bogged down he had to unhitch the sled, dig everything out, and then break-trail with just the snow machine. Then he'd hitch up the sled and drag it along the prepared trail. He had snow shoes which had to be strapped on every time he walked any distance over the snow. They were cumbersome and he couldn't wear them while working around the machine. Consequently, he spent a good part of his time floundering up to his thighs or worse in the snow. Each time he even nudged a tree, it brought an avalanche of snow from its upper branches.

"It was damned hard work, time consuming, and extremely frustrating." Hagen said. "There were times when the machine almost buried itself, and me. Once the front of the machine thrust under a show-hidden tree trunk and became firmly wedged. It took nearly an hour to get that straightened out. The machine never steered the same afterward."

One disaster after another for two days and finally, totally burned out physically, and with his machine bent and scarred, Hagen admitted defeat. Not something that came easily. He returned with considerable difficulty along his trail to the Jeep, loaded everything and drove to Tok thirty-five miles away. He planned to rest in the Motel for the night and beat a despondent retreat back to

Anchorage. It was after a comfortable night's sleep, as he was devouring a huge breakfast in the Motel restaurant, his fortune changed for the better.

He struck up a conversation with a grizzled old-timer named Tony Dale. Hagen was not one to start conversations readily, but Tony's appearance attracted his attention. He looked just like an Alaska frontiersman. At age sixty-five he was big like a bear and sported a bulky, neatly trimmed, beard. "He wore a great wolf fur coat and a fur hat." Hagen said. "Both looked as if they had enough life left in them to give me a hell-of-a-bite if I got too close." Tony turned out to be about as kind and gentle a person as one could find anywhere.

Within minutes, Tony had beguiled Hagen into unloading his tale of woe. Tony relaxed and listened patiently with an occasional grunt. Hagen didn't realize it at first but he could not have met a better person right then. Tony Dale, was a long time resident of Tok. Contrary to his appearance, he was a surprising combination of retired high school teacher, lay preacher, and trapper. He was also something of a local legend. Hagen discovered later Tony had a reputation for helping people who were genuinely in trouble. He obviously recognized Hagen was a true case in need, and without hesitation volunteered to help. It wasn't long before they migrated from the restaurant to Tony's rustic cabin a few miles away.

In his well-equipped workshop, Tony looked over Hagen's battle-scarred snow machine and sled. He immediately pointed out some things they he could easily do to make the machine more capable of negotiating deeper snow. His own machine, standing nearby, was an example and it didn't take an engineer to see where simple modifications had been made. Within a very short time he and Hagen were cutting and trimming strips of metal, welding them securely onto the edges of the snow machine's front skis, and to the outer edges of the sled runners. Tony explained, and even roughly calculated in pounds per square inch of surface, that the existing runners were too narrow to bear the front end weight and ride over soft, dry snow typical of this area. There were other modifications, tungsten carbide-coated metal studs on the snow machine track belt. Those required parts and had to be specially purchased and could not be installed now. They also spent time straightening out and beefing up other parts damaged during Hagen's ordeal.

With the minor changes, Hagen, during a short test run, could tell there was a tremendous difference in his machine performance in soft snow. "This is great. I gotta call Doug." Hagen told Tony.

Hagen and Tony got along so well, Hagen was invited to spend the night at Tony's house and they sat by the log fireplace and chatted long into the evening—kindred spirits despite the difference in age. It was during this evening that Tony divulged his most valuable information to Hagen. Swearing Hagen to secrecy—on a stack of Bibles—he revealed the location of the entrance to his trapline trail which led to the western end of Ladue Valley. It was south of where Hagen had just had so much trouble. Tony explained the snow would still be deep but there would be his own trapline trail, well used, to follow. It added about twenty miles to the round trip, and it entailed crossing the Ladue. Hagen was assured this was no obstacle this time of year. The easier traveling would be well worth the extra mileage.

Tony's help and hospitality gave Hagen his confidence back. He said goodbye to his new friend and set out once again.

Tony's trapline trail head wasn't easy to find. But with a little effort Hagen found it and unloaded his machine. He was soon heading into the woods via the secret entrance. Secret, except that his machine left a pretty clear trail. Tony had been right. Once past the entrance, the trapline trail was fairly defined, easy to follow. Fallen trees had been cut and cleared away where necessary. The modified runners helped enormously in keeping his machine from bogging down. Occasionally he got stuck and it was back to shoving, heaving, and winching. On the whole, however, it was much better and Hagen made good progress toward the Ladue Valley.

Hagen had learned a hard lesson. "Quit," Tony told him. "While you are still feeling good," and, mindful of Tony's advice, quit while he was still feeling good. "Quit while there is still daylight to build a comfortable camp." He now resorted to contemporary measures by unloading a small chain-saw and noisily attacking a fallen tree for a cheery fire. He spent a comfortable night tented by his camp fire. When light allowed, he was up and going again.

It was about ten AM when he finally made it out onto the open, wind swept, Ladue Valley. Here the snow was not deep and the surface was much firmer. There were few trees except near the course of the river. Almost immediately he found himself confronted by the surprisingly high banks of the Ladue River. The river appeared to be completely frozen and then covered by drifted snow. He carefully crossed where the ice was windswept and visible. "You can gauge the depth and safety of river ice, if you can see it." Tony tipped.

Making a mental note of the landscape and features, so he could find the trapline trail again on his return journey, he struck off east along the wide Ladue Valley. What a relief it was to cover several miles at a reasonable speed without getting stuck.

He learned to appreciate depth and distance of winter travel. At low temperatures, nothing was really clear at ground level. The light breeze picked up fine ice crystals which whirled around in the air and created a thin, glittering ice fog. The sun shone through the fog from a low angle and sometimes looked surrounded by a beautiful halo. The outer extremities of the halo, now to the southeast and southwest, created vertical shards of light, some with color. Alaskans called them Sun Dogs.

Thus, Hagen traveled through a surreal landscape with limited views, at least at ground level. With the sun filtering colorfully through the patchy fog, it was beautiful, if little disorienting.

Hagen knew roughly where he was. The tops of the high ridges were visible to his left above the layer of ice fog and trees bordering the Ladue were usually to his right. The position of the sun for the time of day served as a further reference which he used almost instinctively.

He was making good time. Maybe fifteen miles into the valley. Without being aware of it, he drove into a patch of dense swirling fog. Suddenly his snow machine reared violently, skidded sideways, and jackknifed with the heavily loaded sled. For a moment Hagen was hanging on for dear life. The rig came to a slithering standstill.

"What the hell?" He killed the engine, and dismounted. His booted feet went out from under him and he fell flat on his face. He was sliding, sliding, sliding.

"What the—? Damn, Whoa!" Just for a moment he panicked, wondered when his slide would stop and what terrible end was waiting for him. His sliding stopped, and after a few seconds he cautiously, and with some difficulty, got to his feet. Surrounded by swirling vapors and ice fog, the only tangible object he could see was his tangled rig twenty feet away. It was resting on level ground but he found he couldn't walk to it.

Everything looked level in this white out. It was an icy slope with steaming water running down it in little rivulets. Then he noticed the smell. It was rotten, like rotten eggs. Sulfur. Yes, that was it. Sulfur.

What was this Hades kind of place in the middle of what he'd sworn was a flat valley? He stood still for a moment in the cold, swirling fog to collect his wits.

Hagen had been stopped by a warm spring where water rose to the surface, spread out, and eventually froze. That was it. As the water froze, it built up an ice dome which would eventually reach fair proportions toward the end of winter. Five months of freezing temperatures would see to that. Of course, the warm and obviously sulfurous water, clashing with the frigid air, also caused the heavy vapors. He had run his machine into the fog bank and straight up the sloping side of the ice dome.

Now that he had it figured out, Hagen set to work recovering his machine. His snowmachine boots did not grip on the wet, glassy, ice and after a fall or two, he crawled the twenty feet up the slope to check the machine. It didn't seem to be seriously damaged. Just a few more dents and dings.

"Just how big is this ice dome anyway?" He asked himself.

He crawled up a bit further and reached the top. Water, terrible, fizzy, smelly, water, bubbled to the surface and trickled down the slopes. Yuck!

"This was not a good source of drinking water." He whispered jokingly. The ice dome was close to ten feet high, above the surrounding flat terrain.

The mystery solved, Hagen inched his way back down to the machine, mounted it, started the engine, and, with some slipping and sliding, eased the rig down the icy slope to level ground. Orienting himself, by the sunlight streaking through the mist from the south, he drove carefully away from the mound and soon found himself back out where there was some visibility. He again identified his surroundings. He coaxed the snow machine to cruising speed again and headed east.

A trouble free two hours put Hagen at the southern end of the Trib 1 Valley at the confluence of the stream and the Ladue River. The valley from which the stream drained, extended northward about three miles from the Ladue River. Hagen's goal was the head of the valley and The Glory Hole.

The driving force behind Hagen's trip was, of course, gold. Hagen and I had hiked into Ladue the previous summer to prospect. We found color in the stream now named Trib 1. It was now a registered claim. Hagen's idea was to ferry in additional tools and equipment with which to work the claim next summer. I used my Cessna 150 to drop supplies into the valley in preparation for a big hike. The Cessna's capacity was limited and some things, like sluice-box and planks, simply would not fit. Hagen thought a trip by snow machine, with freight-sled, was the answer. Here he was, forty miles in from the nearest highway, wet up to the knees, and with racking back pain.

It was at the confluence of Trib 1 and Ladue where Hagen hit big trouble. He didn't realize the danger until it was too late. Under a thin cover of snow, water from the Ladue overflowed the surface, in a sizable area, then froze. The machine crashed through the weak upper surface and came to a steaming halt on the original layer of ice a foot under slush and water.

A hell of a way to learn about mother nature. It took him an hour and a half to hook up a long winch rope and drag the machine, then the sled onto a firm surface.

A trail of jagged holes bubbled, where the machine and sled dropped through the first thin layer of forming ice. An hour of floundering around to recover the equipment, left his feet and lower legs wet and ice-caked. At first the insulated snow machine boots acted like a divers wet suit and Hagen's circulation helped to maintain a safe warmth. Now he was out of the water and the bitter arctic cold began to penetrate.

To make matters worse, Hagen had slipped on the ice and wrenched his back. Pain and muscle spasm plagued his body. Still, he had to keep going and quickly get himself dry and warm or he'd have frost bite.

With his machine on a secure surface, he painfully limped to the clump of trees where he had anchored the winch rope, collected a few dead branches, shook off the hoar frost, and set a fire.

"To hell with safety." He said. He dribbled some gasoline on it, stood back and threw a match. Woomph, a one match fire, as good as any boy scout could have made.

From his kit on the sled, he unpacked a spare set of clothes and boots, and went to change from his wet things. For a couple of minutes he stood almost naked in the freezing air. As fast as the stabbing pain in his back allowed, he shivered, shook, and cursed his way into dry thermal underwear and trousers and finally zipped up his one piece snowmobile suit. By this time his fire was established and he was able to set about preparing hot chocolate.

"Give me warmth," he pleaded. He rigged up the winch line and hung his wet clothes in the radiating warmth of the fire.

Hagen's will dwindled and he was brought, almost, to tears by his situation and his pain. He wrapped a tarp behind himself as an additional wind break and hugged the warmth of the fire.

"Wish I knew whether to go on," he said through chattering teeth. Getting bogged down so frequently was frustrating but not very dangerous. The incident with the ice dome, was scary, but fascinating. This water hazard, however, downright dangerous, was a killer. He would have to stay away from the river and stream as much as possible. He may not be so lucky next time.

He had been lucky the trees had been close enough to use as an anchor for the winch rope. The machines were clear of the water. He was lucky the water was not deeper, and that the second layer of ice held. He was lucky the materials were on hand for a fire. Now he was in dry clothes and beginning to warm, things looked a little brighter. The only damage was his wrenched back, a shattered ego, and a few more dings on the snow machine.

He stood and rotated experimentally. The pain made him cry out. His breath came in short gasps. His back injury was not going to make things easier.

It was after one o'clock already. With all the excitement, Hagen almost forgot the rendezvous with me, in the air. "Doug should be here any time soon." He said hopefully. He delved into his kit and fished out the radio, checked it, then stuffed it inside his snow machine suit. A warm radio phone worked better. "Lunch. I've got to have some food." And he dug food from his sled.

Forty-five minutes later he was savoring a second cup of hot chocolate when he heard the distant aircraft engine. My timing was right as I flew east in the winter-cold. "Sure enough," Hagen told me later. "There was a small speck cruising at low altitude along the course of the Ladue. I wondered if his snow machine trail was visible from the air. Maybe, I thought, it had already been obscured by blowing snow."

I followed the Ladue until Hagen could see my red and white Cessna 150. "A little late, but such a welcome sight." He said. "I was so relieved and so caught up with excitement I almost forgot I was supposed to use the radio to report my position."

I was just south of his location and flying through a sun dog - when he first keyed his radio and sent out, what seemed like a very lonely call.

"Come in, Huh, Come in Doug. Over?" Silence. "Come in Doug. Over?"

Suddenly, the radio squawked to life. A lot of background static, but not too bad. "Hagen. You made it. Where are you? Over." His voice sounded strange in that vast expanse.

Hagen was elated. "Doug. Doug. Hell, it's good to hear your voice. I'm on the west side of Trib 1. You're at Ten o'clock."

Boy, that sounded professional. I tilted the wings of the Cessna so Hagen knew his message had been received. In took only a few seconds for me to zero in on him and pass noisily just to the east of his position. "Okay I've got you. Is everything all right?" I banked the plane into a wide left turn at two hundred feet and circled out over Trib 1 Valley away from Hagen.

"Yeah I'm okay. Well, sort of okay. Got my feet wet here but I'm drying out. Gave myself a crick in the back too, but I think I can work it out. Could you see my trail? Over?"

Screech. Screech. "Damn, getting out of range already. This radio really has a short range. Maybe the battery is cold. Never mind, coming around again. Over?"

I circled his position again hopefully into radio range. Finally. "Guess you didn't hear me, Hagen. . . . I can see your trail in some places. How did it go? Over."

"Like I said when I called you from Tok, the first attempt was a disaster. Didn't think I would make it. Tony was really helpful. Thanks to him the machine handles much better.

Still a few problems through the trees. Are you receiving me?" Click, click.

"I hear you."

"Okay. Really good traveling from the trees till I hit the river water here. Terrible trouble. Oh! There was one strange thing, but I'll tell you about it later. Good thing the water wasn't any deeper... Machine looks okay but I haven't tried to start it yet... Too busy drying out and warming up. Hang on for a minute."

Hagen primed the snow machine and pulled the starter cord a few times. The engine puttered and died but the third time, it started and after some hesitation settled down to a smoky, fast idle. Stepping away from the engine noise Hagen keyed the radio again.

"Yeah Doug, it seems okay. Over?"

"Hey, that's great. I can see the holes you made. That area looks gray compared with the rest of the snow. I can see quite a few similar patches by the Ladue and a couple of a hundred yards north on Trib 1, so stay well to the west. Sorry I was late. Problem getting out of the tie down area. We had about six inches of snow last night and nothing was ploughed. The flight itself was fine though. Great. Stand by, I'll drop a package. Read me? Hagen."

"Yeah, I read you. Make the drop west of my camp. I don't want to go wading again."

Hagen stood in one spot and turned to keep the plane in sight. I made a wider loop, lined up with his camp and descended to a lower altitude. From Hagen's perspective it looked as if I was going to touch down right where he was standing. I partly extended flaps and slowed. The sunlight flickered off the propeller.

Hagen recalled later. "I saw the pilot's side window hinge out and upward. Just before it reached me, the plane banked just a little and passed to my right. The small, light weight package popped from the plane and landed in the snow about fifty yards away. 'Excellent drop, Doug. Excellent drop.'"

Hagen, painfully, retrieved the package then returned to the fire. I circled, staying within radio range and watched. Quickly Hagen opened the string-wrapped package. Dry thermal socks, thermal underwear, a wool shirt and two big bars of chocolate. He held up the chocolate in a thumbs-up salute.

"Well, I said to myself, "he wasn't desperate for such items, but one never could tell. It was the thought that counts."

Keying the radio, Hagen almost shouted. "Good timing Doug. I just used my change of clothes and the wet things are still wet . . . Huh, Damn! Correction, frozen stiff. I'll manage better once I've unloaded the sled, much lighter on the return. I'll stick to the same trail . . . except the lake here. Shouldn't be too much of a problem You'd better get out of here or it'll be dark before you make it back to Anchorage. Over?"

"Right . . . I don't have too much time. If I have to, I'll stop at your place tonight. Glad I was able to locate you. Air temp. shows -28° so be careful down there. Don't take unnecessary risks. Careful with that back, too. Are you sure you'll be okay? Over?"

"I'll be okay. It was tough going, but I don't think there's anything worse than what I've already been through. Thanks for the company. Oh, and the package. Take care going back. Over?"

"See ya back in town Hagen. Be careful. Over and out."

I passed his camp as noisily as I could as a goodbye signal. I saw his hand waving briefly as I gained altitude. I headed for Tanacross.

Hagen knew I would refuel there and then fly directly to Anchorage and arrive just before it was dark. That was a lot of flying in one day. "If Doug is pushed for time he could drop in at Wasilla." Hagen said aloud to the lengthening shadows. "Short of Anchorage, but he knows where to find the house key."

The brief rendezvous was over. Hagen was left with the company of the puttering snow machine. Satisfied it was okay, he hitched up the sled and drove the rig closer to the fire, then switched off the engine. A profound silence descended over the area. "Minus 28°, huh. Thanks Doug, I might just need that underwear and wool shirt." He said. As an after thought, he pulled a flat Kodak disc camera out of his pack and snapped a couple of photos of the trail, where he had broken through the stream. It was already freezing over.

"There is less than three hours of decent daylight left. Time to break camp and make it up to Trib 1 claim before making camp for the night. There were always many things to do in preparing for a night in such low temperatures and everything took twice as long as normal even without a painful back. Am I talking to myself?" Hagen shook his head in dismay.

In a few minutes Hagen was on the move, gritting his teeth against jarring pain. He was now in familiar territory, having hiked with me through this part of the valley last summer. On his right was the place we made camp the first night. A little further north stood the sheer rocky cliff. With the snow cover disguising many features, he failed to recognize the small escarpment where the stream dropped a few feet and we first panned for gold. The machine was airborne as it mounted the ramp. "Ouch! That hurt." He gave the stream a very wide berth in case there was another overflow.

Toward the head of the valley there were many tamarack and spruce trees and a tangle of dwarf willows. This made hiking difficult last summer. Now, with snow covering most of the lower growth, traveling was comparatively easy. Hagen kept the speed up and, in spite of deepening snow, kept moving without incident.

The Glory Hole. Hagen recognized it only from the lay of the land, one particularly prominent outcrop of rock and a large tree trunk resting like a bridge over the stream. Everything else was covered with a feature-softening, snowy blanket. Sitka spruce on the east bank stood guard like great white-draped sentinels. There was no sign of the stream which passed by the ledge of rock, and, when Hagen switched off the engine, not even a sound of running water. Playing it safe, he donned his snowshoes and probed the area with a long stick before driving over the stream bed. He knew where he was heading. On the rising tree-clad ground to the southeast of the claim, there was a particular rocky outcrop behind some spruce trees. We stashed tools and left over supplies the summer before. "Like an investment for the future," he thought. He'd unload the sled into the same cache tomorrow. Right now, he wanted to set up camp and protect the snow machine before darkness fell.

Aware the snowmachine left a defined trail, he didn't stop by the cache. Instead he went barreling across the slope a short distance and stopped at a little plateau where the snow had been blown away. This spot was also sheltered by a stand of larger conifers. "A good place for a warm camp." He said.

Unlike summer hikes and camping trips, Hagen had to use a tent for shelter. It was just a four-seasons dome tent, which set up very quickly and needed no stakes but offered protection against cold and wind. Inside was a porous fabric and allowed moist inner air to escape. The outer shell stopped the wind. If any ice formed due to condensation, it did so on the outer shell where it could be easily beaten off.

Hagen busied himself with the camp. First the snow machine, it needed to be protected from the worst of the cold or it might not start the next day. A small oil-fired heater was set alongside the engine, then the engine compartment was completely covered by an insulated blanket. Treatment that was common for planes, and snow machines, in such subzero conditions. The heater would burn gently all night giving enough warmth to keep the worst chill away.

Next came the tent. He shoveled a level patch in the two feet of snow and set up the tent with the edge flaps pinned down by snow. The kit bag, rolled ground pad, and two bulky sleeping bags were placed along the inside edge. The sun set behind the shoulder of the ridge as the campsite was secure. The air was much colder already. If it was -28° in the sun, it could easily get 20° colder during the night.

Next priority was a fire. With the chain-saw he soon reduced a nearby fallen tree to manageable logs. The smaller branches would be used as kindling. Hagen soon had his fire glowing and crackling. Even a small amount of radiated warmth felt marvelous. Every survival book we ever read said the same thing. "Make a fire and set up camp as best you can, even if the stay is short." He rigged up a rack from which he hung the still frozen clothes. They would freeze-dry if nothing else.

Hagen endured the twinges in his back, but with the work, it began easing a little. He had to be very careful to keep his fingers from freezing. Some things just could not be done while wearing bulky mittens. He removed them for short periods, then quickly replaced them to prevent frostbite. Everything took much longer, and had to be done with care, under these arduous conditions.

Eventually, Hagen cooked his dinner and then huddled close by the fire to eat. There was no thought given to any kind of ablutions. Even a necessary call of nature was an uncomfortable experience. Alaskans often used the expression, "Colder than a well diggers butt," and Hagen related with the truth of the saying.

It was quite late when he stoked up the fire one last time and struggled into the confines of the tent. It was a work of art to get out of the bulky snowmobile suit and boots and into the tent without dragging a lot of snow inside. He was getting better at it but it was still a relief to snuggle up in the doubled sleeping bags. It would be a long night, but what else could he do but sleep, if he could. "Hope my back feels better in the morning." He groaned.

It was still dark when Hagen awoke on the morning of the fourth day, a call of nature prevailed. While he was up, he dropped a few small twigs on the embers of last night's fire and stirred it to life. It was so cold his eyelashes froze together when he blinked. It must have gone down to -40°

during the night. An ice fog was settled over the camp and in the firelight the thick coating of frost shimmered and sparkled. Hagen decided it was too early to contemplate staying up and freezing. He placed sizable logs on the fire and retreated again to the warmth of his sleeping bag. Two or three more hours sleep would feel pretty good.

Nine o'clock saw him out and about again. It was getting light now and he could see to move around. The fog still hung over everything but it was thin and would probably clear as the sun came up.

Hunched close by the fire, Hagen had a breakfast of hot oatmeal and raisins, rye bread and cheese, and a pot of hot coffee which, made from melted snow, tasted blah. Today he would stash away the sluice box and other equipment and then head off home. He felt confident that traveling out would be easier than coming in had been. Besides, his back, apart from an occasional twinge, didn't feel too bad this morning.

"Time to go to work." He donned his snowshoes and struggled through the deep snow to the rock between the conifers and began to dig with his aluminum shovel. Hidden away in a small cave, like space on the downhill-side, were the tools which he and I had air-dropped last year. After using them for initial prospecting, we stashed them away and covered them with sturdy logs. The whole cache was now under four feet of snow.

It took sometime to dig down and the logs made it difficult to clear a work space but eventually the cache lay exposed. Everything was intact which was a relief. Even in this remote place, trappers, hunters, or gold prospectors, may have seen signs which led them to the cache.

Hagen pulled the logs away and then cleared a space. With difficulty, he carried the items from the sled and stacked them neatly under the rock using the planks of wood to shelter everything. With all the items cached, he carefully replaced the protective barrier of logs and then shoveled the snow back into the excavation. His activity left a defined trail to the cache, so to help obliterate the evidence, he used a large pine bough like a broom, and smoothed his snowshoe tracks. The next wind or snow fall would blend the snow surface. The cache would not be visible.

Good old standby—beef and vegetable stew sufficed for lunch, especially followed by hot sweet tea. After lunch, Hagen topped up the snowmachines fuel tank from a spare can, broke camp, stowed his belongings securely on the sled, and readied himself for the journey out. Thanks to the heater, the snow machine started straight away and he let it run for a few minutes before moving off.

The sled in tow was now much lighter and Hagen's spirits lifted. The machine responded quickly and was much less inclined to bog down. Plus, he had a trail to follow. "Now, this is more like recreational snowmachining and less like work." Hagen grinned. He soon reached the place where I made the air drop, and mindful of his earlier experience, stayed well away from the hazardous spot near the stream. It was time to turn west, so he picked up remnants of his original trail and coaxed the machine up to a comfortable speed of fifteen miles an hour. There was no point in blazing a new trail when he knew this one served the purpose.

This time he was ready for the ice dome, and spotting the thickening patch of fog, he made a slight detour to the south, and then picked up his original trail in a mile or so.

By four o'clock, he reached the Ladue, crossed with care at the same place, and reached the trees at the western head of the valley. It was too late now to think of entering the trail through the trees, so he would have to set up camp for yet another night. It was still bitter cold, but here, among the trees, there was no wind and felt warmer. Once he had his camp set up, and a fire burning brightly, it was quite homey. Before it got dark, Hagen started the chain saw and made short work of a fallen tree. With a good supply of firewood, he prepared his dinner—another can of beef stew. "Much more of this and I will look like a can of beef stew." He mumbled as he scraped the last of the can into a kettle. He flattened the can and added it to his stash. Most were labeled Beef Stew.

Nighttime in the wilds was a time of deep contemplation for Hagen. It was icy cold, but with hot tasty food, endless hot drink, chocolate or coffee, and a bright fire, there was a certain excitement to it all. In Hagen's mind—although he had a snow machine and other modern conveniences—he was following the early mountain men, prospectors, and trappers who preceded the settlers in the later part of the nineteenth century. All sorts of things went through his mind during the hours spent hunched in the small circle of warmth listening to the hissing and popping of the flames and watching sparks drift into the starlit sky. He was steeped in reverie and it would have detracted from the atmosphere if there had been anyone with him.

A couple of times bright eyes sparkled outside the ring of firelight but they were small, possibly those of a fox. It had crossed his mind there might be wolves around, but he had his rifle nearby. He was also thinking of the energy expended on this trip. It seemed like so much work, not to mention danger, just to deposit a few tools and equipment on the claim. He knew it would pay dividends when we hiked in there next summer. With a good-sized, advanced air drop of food we should have an easy hike and a couple of productive weeks on the claim. We already invested quite a bit of effort, time, and money in this venture. Maybe this year we'd see a return on the investment. Later we would have to find a vehicle to make the journey. Backpacking was time consuming, and eventually, if the claim proved to be worth it, we would need mobility and more equipment to do the job right.

Tomorrow was another day and Hagen was confident that it would not be too difficult. With a good start in the morning he should be at the Jeep before noon. Lunch would be at the restaurant in Tok. Perhaps he would invite Tony and he'd tell him how things turned out. He surely hoped they were serving something other than beef stew for lunch. In the afternoon he would be wheeling toward home in Wasilla.

He spent the night quite comfortably and was up and about at first light. A hot cereal breakfast, a little while packing everything securely on the sled, and he was set to move out.

Now to see if the return journey really would be better. He was under no illusions, even with the lightly loaded sled, it would still be challenging. He slipped his goggles down, fired up the snow machine, and hit the trail. He left Ladue Valley and headed into abrupt changes. The terrain rose gradually and it was heavily forested. At the lower level, trees were skinny, shallow rooted, arctic tamarack, and black spruce so prolific in wet tundra areas. At higher levels, trees were larger, and comprised of evergreens and deciduous. All of the evergreen trees were sagging under great weights of snow. The deciduous trees stood stark against the sky, their naked branches coated with frost. Occasional clearings were partly blocked by exposed upper branches of dwarf willows.

He stuck to Tony's blazed trapline trail as much as possible, noting, with some satisfaction, that his machine negotiated the trail much better with the lighter sled. Also, he supposed, he was a little more proficient too. He seemed to sail straight through some of his former trouble spots. Once he overshot a turn and the machine and the sled took a nose dive into deep snow. Oh well, back to the shoveling, heaving, and hauling. Thank goodness his sore back was feeling better and didn't stop him from doing such things.

He made it to the end of the trail and the waiting Jeep just as he had planned before noon. Loading the sled and snow machine on the trailer, he soon had the Jeep rolling along the AL-Can toward Tok. It had been tough but the mission was a success. Another determined stride in our mining ventures, and most importantly in his own mind, another challenge won.

# Chapter 20
## Working Gold Claim

It was the last week of August. The aspen and silver birch trees in the Trib I Valley were in full color, contrasting sharply with the dark Sitka spruce and other conifers on the surrounding hills. At two o'clock in the afternoon it was mild with temperatures hovering around 65°, cooling off rapidly in the later hours.

The valley was changed. Until lately, it had been a peaceful place with only the sounds of nature breaking the silence. Now, the silence was broken. Man and his technology intruded—the beat of a five-horsepower gasoline engine, the raucous sound reverberated off the hillsides. It was difficult to pinpoint the real source of the sound.

The Glory Hole was the center of activity. Located at the narrowing head of the valley at a point referred to on maps and registered at the mining office in Anchorage, was shown as Trib 1, H and D Claim. Named after us, the owners, Hagen and me, Doug.

We were busy on the claim but other things kept us busy too. Our claim area showed evidence of earlier development and activity. Carefully, we had not disrupted the land more than necessary. As we reconnoitered, it was obvious much of the natural landscape was disturbed.

Hagen and I set up a substantial shelter, a twelve foot by twelve foot army tent, reinforced with a sturdy plastic tarpaulin fly-sheet. It stood on raised ground to the east of our claim. By the west side of the stream, some fifty yards from the "cabin," was our large excavation. To the side of the excavation was an elaborate sluice box and a hopper system. The engine driven pumping unit, placed near the stream, was busy pushing quantities of water through a two-inch pipe connected to the upper end of the sluice box.

Other equipment stood around our site, including a small portable electric generator, a gold panning machine which we used almost every evening, and a suction dredge nozzle. Our buckets, shovels, wheel barrow, and smaller tools were on site. Some comforts of home were hidden inside the shelter.

We had ferried equipment to Trib 1 over the last two years. Our equipment was dropped from the plane, hauled in by snowmachine during the winter, and most recently by a sturdy six wheeled ATV nicknamed Herman. The camouflage painted vehicle presently stood alongside the cabin.

Two years had passed since Hagen and I had made the initial exploratory hike into the Trib Valleys. After that hike, and another last year, we were convinced this particular site was worth developing. We purchased the ATV, as a stock vehicle, from a dealer in town. During the winter, we made extensive modifications and turned it into the kind of vehicle we could really use. An eighteen-inch extension was attached to the general length of the chassis. The box bed, a second twelve-volt battery, a fifteen-gallon reserve gas tank, an electric winch, a sturdy roll bar, front bumper, and finally a camo paint job rounded out the project. We added a powerful swivel light, and loops for lashing down cargo.

We also purchased a second hand tandem-axle boat trailer and adapted it as a transporter for the ATV. It was stashed at the trail head along with the old Chevy Blazer.

The trip to our claim was a breeze on the ATV, specially compared to the tough hiking we did in the past. Of course, it simplified the ferrying of supplies. It was impossible to do on foot, or by snowmachine, and sled as Hagen dared eighteen months ago. With help of the ATV, fresh food, not considered while hiking, was hauled in to camp. We gladly added, onions, potatoes, and carrots to the pot.

With the big, six, nobby tires mounted on fifteen inch by twelve inch rims, the ATV had beaten a pretty well-defined trail from the main ridge trail down into the Trib 1 Valley. Still, we were protective of our claim and took pains to disguise the entrance from anyone passing the main trail. We had never yet met anyone on the trail but we now knew it was occasionally used by fall hunters.

Our Trib I claim turned out to be a good producer and even more so since we constructed a larger capacity sluice box and hopper. It required a little adaptation, but thanks to innovative valving and pipe connections, the arrangement could be converted into a suction dredge able to vacuum fine materials from depths of our excavations. It had proven to be a big advantage when the material was suitable and the hole was filled with water. This year, by consolidating vacation time, we spent almost four weeks on the claim. Now, we were just starting the fourth week and would have to pack up in a few days.

Throughout our trips to the valley, we had never seen a grizzly bear. Much of our trepidation disappeared. We did see an occasional black bear and even had one which regularly snooped camp during the night. We kept our guns handy. Hagen's pump shotgun was propped against a nearby tree and I sported a 357 magnum in a shoulder holster. It was a bit awkward at times but I felt it was worthwhile to have it handy. The chattering noise of the small engine was something a bear, especially a hungry one, might get used to after a while.

At three o'clock, we started one of our regular cleanups of the sluice box. We raked out the larger rocks and gravel, lifted up the riffle assembly and removed a snug fitting strip of indoor-outdoor carpet from the bottom of the box. The carpet, along with its load of concentrate was rolled and carefully placed in a bucket of water. Later, we'd rinse it thoroughly in a larger galvanized tub.

This was an exciting time. If we sluiced good sized nuggets, we'd see them in with any larger flakes as the riffles were lifted. Much of the finer gold remained hidden deep between the coarse fibers of the carpet rinsed out later. We usually idled the engine and worked on cleaning the sluice. It was a relief from the hard work we shared, but we also shared in the discovery gleanings from the earth. At afternoon tea break, we shut down the engine completely and let our ears rest too.

Clean up took us about half an hour. There were no record breaking nuggets. After three weeks, we were past great excitement over small nuggets. We realized, however, every little nugget or flake helped to increase the total poke. In the evening we rinsed the fine gold from the pieces of carpet and set the panning machine going to separate the gold from the heavy black sand. Once the sluice was all set to go again, with clean carpet installed, we went across to the cabin and brewed tea.

Breaks were much easier now. We had a single burner naphtha stove we used for boiling enough water for tea or coffee. Our main cooking was still done over a small fire. We occasionally heated a large galvanized tub of water for laundry or a hot shower. Yes, hot shower! Enough of cold water, hardy, frontiersmen nonsense. We had cleverly rigged garden hose so the water circulation pump of the panning machine would suck up the warm water and spray it from a shower head set above the tub. It was marvelous therapy after a hard day of grubbing in the Glory Hole.

For rainy days, and there were a few, we had a small potbellied wood stove set up inside the cabin so we cooked using a combination of the naphtha and the wood stove.

With the pump shut down, and the claim area quiet for the time being, we sat on a couple of aluminum garden chairs and waited for the water to boil. From this elevated vantage point, we surveyed our claim and the hills to the west. When we arrived three weeks earlier, the hills were brilliant scarlet with fireweed. In these three weeks we watched the blooms of the fireweed march progressively higher along their stems. When the flowers were at the very top of the stems, autumn began. Other flowers were now overwhelmed by Autumn yellow and orange colors of the deciduous trees.

Last year we started digging and sluicing below the cabin on the west side of the stream. A narrow trench about twenty yards in length was left. Probing further with a long crowbar, we were extremely fortunate to identify a reef of bedrock about two feet below the surface. The material against this impermeable reef had been the most productive. With our improved machinery, we were exploiting this trench to the fullest, and consequently had branched out to follow the bedrock.

This year, with the larger sluice box and suction dredge attachments, we continued on a broad front. We used the wheelbarrow to bring the overburden material to the hopper end of the sluice box. We now had a large excavation and allowed it to partially fill with water. It would then be suctioned using the dredging nozzle. Recently we began to back fill the hole with the tailings from the sluice so that it virtually remained a constant size more easily flooded with water when necessary.

When we started the sluice the previous year, we used the tailings to form a settling pond so the fine sediment would not foul the stream. We were careful with the pond and regularly checked downstream to gauge its efficiency. Our pond worked well and our mining activity had negligible impact on the environment outside the immediate claim area.

Getting to the sluice from the cabin involved crossing the stream, so we positioned sturdy tree trunks as a bridge. Our camp was well arranged and comfortable. There were times when we felt more like two of the seven dwarfs trudging off to work in the mine every day. Hi ho, hi ho.

For rest and recuperation we found fishing in the Ladue River to be excellent. Occasionally, we would take a day off and ride the ATV three miles south to the river where we would fly fish for grayling. Hagen was not keen about fishing, but after a few trips, he was hooked and really enjoyed using a light fly rod.

The Ladue River ran a dark tea color from peat leaching from the swampy areas of the wide valley. I heard there were medicinal benefits from swimming in such water. True or not, on the nicer days, to test the theory, we braved the cold water, and skinny dipped. We were never sick, so we figured the tale must be true.

Time on the claim each year was a premium, but a month was a long time to spend processing dirt through the sluice box. We fell into a pattern of spending every fourth or fifth day relaxing. We sometimes hiked a little way up the hillside and picnicked at a higher elevation and enjoyed the good view of the surrounding countryside. This served as a great escape from the claustrophobic confines of the narrow valley, and gave us a different kind of activity.

Last year, during one such hike, we discovered something strange. About half way up the wooded slope behind our camp, we came upon three large circular holes. Each was about forty feet across and filled with water within a couple of feet of the rim. There was no telling how deep they were. Part way up a thirty-degree slope was a strange place to find such ponds. The only thing we could think was that they were bomb craters. Perhaps in earlier years the military had used the area for bombing practice. Anyway, if that's what they were, it was a long time ago because the holes were now surrounded by mature trees.

Blueberries were in abundance at the higher elevations and we sometimes picked a pint or two for dessert. An occasional grayling and the blueberries were easy variety to our diet.

Well, tea break was over, so it was time to get back to our mining job. We crossed the bridge, pulled the engine back to life, and started sluicing. Hagen was whistling, "Hi ho, hi ho, it's off to work we go."

It was my turn to bring dirt to the hopper and Hagen's to feed it bit by bit into the sluice. We rigged the hopper so that a quantity of material could be dumped in and then metered into the rush of the pumped water. Everything cascaded over a half inch mesh screen and then over the series of twelve riffles in the long sluice box. It was important not to feed the material too fast or the riffles could clog. It was equally important to make sure each rock was carefully washed because fine gold tends to stick to the clay on rock surfaces. It was a combination of water force and gravity. Water, to wash the material and separate grains of clay, sand, gravel, and gold, and sweep them down the sluice. Gravity, to cause the heavier elements, called concentrate, including gold, to fall to the lowest point between the riffles where it could not be washed away by the force of the water. We always used sturdy rubber gloves when tending the sluice because there was a lot of hand work and the water was icy cold.

Suction dredging the depths of the excavation was, in its way, a little easier for us. The size of the nozzle ensured only smaller gravel and fines were sucked up and there was little washing necessary. We proved the greater percentage of gold was found at bedrock and suction dredging was the most efficient method of recovery.

Our sluice box spillway carried tailings away from the sluice itself. Occasionally, it was necessary to truck these tailings away and backfill the hole. Thus, the material was moved twice and the excavation never really got any larger.

Experience proved that a piece of carpet under the riffles would last about three hours. Any longer and it would choke with heavy black sand and be less effective in snaring fine particles of gold. Since we didn't want to lose any gold, we usually worked three or four carpet shifts during a mining day. Counting cleanup time, it was almost a fourteen-hour day.

That was the hard work. Our day didn't end there. Every evening we had half a bucket of heavy concentrate which had to be passed through the panning machine. The panning machine was a clever device with an eighteen-inch diameter rotating disc set at forty five degree angle. The disc features a three-inch rim and has a spiral groove formed on its flat inner surface. A little pump delivers water through a nozzle which keeps the disc irrigated. As the disc rotates, there is a panning action which causes heavy gold to settle to the bottom where it is picked up by the spiral groove and transported upward to a hole in the center of the disc. A tube leads from the hole to a container under the machine. The water pump and disc are driven by a small gasoline engine.

The little engine was noisy but we often troweled in concentrate while we were preparing the evening meal, and even little by little while we ate. While we fed it, the machine did the rest and it was fun to watch a procession of gold marching up the spiral and into the container. It was even more exciting when there was a nugget heavy enough to make an audible klunk as it hit the container.

After panning was finished and the gold was dry, we shook it through a special sieve which graded the gold by size into four categories. The gold was then placed in screw-top plastic bottles. The bottles were placed in a metal box, and with typical care, hidden where only we could reach. The box was satisfyingly heavy with results of our three weeks of hard labor.

At six in the evening, we completed our final cleanup for the day and carried the heavy pail of concentrate to the camp and started the panning process. Then we started our other chores and only occasionally paused to ladle in a quantity of new material.

It was our usual practice to clean up first and then prepare dinner. We got into a routine and things usually didn't take too long. Normally, we had everything done around eight o'clock and we quietly relaxed by the fire.

For evening entertainment we had a transistor radio powered from the ATV battery. We picked up amazingly far away stations at night. We often pulled in European stations and obtained a different slant on the news. Unfortunately, there were news items, occasionally, which set Hagen off, especially if it involved politics. Thank goodness there were music and comedy too.

Our cabin was equipped with a few comforts. For one thing we had military cots and sleeping bags. We debated whether we would have been warmer on the ground, but at least we were off the dampness. We also had a couple of foot lockers for clothes, and of course, the small potbellied wood stove. The stove pipe went up a little way and then passed through a fireproof panel on the back wall of the tent. The stove was small but it took the chill from cold nights and helped keep our clothes dry.

To keep the cabin cleaner, we constructed a partial floor made of a series of wooden slats, with half inch spaces. It was easy to brush off occasionally. We even had electric lights, powered by the generator, but we hardly used them. In fact, we decided to take the generator out with us at the end of this trip and not bother with it again. Mostly, if we needed additional light, we used a couple of small lights strung from batteries of the ATV. We were careful not to discharge the ATV batteries too far.

Outside, to the left of the tent entrance, sheltered by the fly sheet which extended beyond the front of the tent, was our cooking table, and on the right side, a strong aluminum chest which held most of our food. This chest was specifically purchased because it was considered bear proof with strong clamps holding the lid closed.

It was beginning to get cold at night. Down in the valley the sunshine faded early because of the ridge to the west. The change in temperature was immediately apparent. By the time our chores were finished, and we could relax, it was cool enough to put on our jackets and huddle closer by the fire for warmth. When it was cool, there were no mosquitoes. Sometimes it was just more comfortable to stoke up the little stove and roll up in our sleeping bags, even if we didn't go to sleep right away.

Tonight we settled at ten. The air was very cool and there was a definite feeling of autumn. Trib 1 soothed us to sleep with soft babbling. A half-moon cast a silvery light over our camp. Unseen and unheard by us, our neighborhood blackie inspected the claim to see how much progress we made during the day.

# Chapter 21
## Day of Relaxation

One day a week was designated as a rest day—well deserved. We never planned because it always depended on the weather. This rest day started cool but promised to be sunny. We decided to go fishing on the Ladue. It was, probably our last chance before closing down the operation for the year.

After a leisurely breakfast, we packed food and drink, rifles, rubber boots, fishing gear, toiletry, and towels onto the ATV. Then we spent a few minutes tidying up the camp and making sure all of our food was locked securely in the chest. We closed the front of the tent as a deterrent to our camp bear, knowing that it could very easily force an entry if it wanted. It seemed to us there was little inside to attract a bear. But, who can tell about bears.

Our frequent trips to Ladue left a clear trail, easy to spot from the air. It couldn't be helped. By this time we were certain a few people knew, by one means or another, of our mining activity in the Trib 1 Valley. Once last year, and again this year, a small aircraft buzzed the area and we felt sure the occupants were scouting to see what we were doing. We never had visitors, not that we knew of, but we were surprised that no one else had prospected and staked a claim nearby. We thought perhaps it was only the logistics of getting there that had deterred them. Perhaps they realized we had the prime spot covered. Not much we could do about it except to be sure we had our claim correctly registered. Anyway, as far as making tracks was concerned, we couldn't stay penned up on the claim four weeks just for the sake of secrecy.

We followed our usual trail south to the junction of Trib 1 and the Ladue River. The place was where Hagen, his snowmachine, and sled, fell through the ice eighteen months earlier. Here the clear water of our stream remained separate for a short distance before blending with the tea colored water of the Ladue. It was our favorite place for grayling fishing and I believe we caught some fish several times. Most of the time we returned them gently to the river, only occasionally keeping one to supplement our camp food. We used very light tackle and barbless hooks, challenging to us, yet the fish were virtually unharmed.

The serpentine Ladue was fringed for most of its length by tamarack and spruce trees. Some of them were large, especially close by the river where the soil was better drained. Generally speaking however, the rest of the wide valley was without growth; just too boggy.

As usual, we parked the camouflaged ATV under the trees. It would be less obvious to anyone flying over the area. By now, the middle of the morning, the temperature was creeping higher and it was beginning to feel pleasant.

We decided to get our lines wet right away and were soon both thrashing water, as I liked to call it. Hagen was challenging himself with forty foot casts to the furthest side where the water ran dark and deep under an eroded bank. I waded into the livelier water where the clear stream and dark river water melded together. I was just downstream of the gravel bar formed by the junction.

Hagen caught the first grayling, a beauty, about fourteen inches long. It was definitely a keeper but we had an unwritten rule not to keep a fish so early in the day. By chance, Hagen caught two more before my first nibble. We had a good laugh when my first catch turned out to be only about six inches long.

Hagen joked, "You're holding your mouth wrong."

"Oh," I bantered. "You're giving me advice now when just a day or so back, you didn't want to try fishing. What's wrong with this picture? I got you hooked on fishing, remember?"

"Doug, I had to take lessons to undo those things you taught me. I keep thinking you taught me wrong on purpose. Do you deny it?"

To take the pressure off I pointed out the gaggle of geese, perhaps a hundred or more, flying overhead. We could hear them calling to each other as they flew in three distinct V formations.

"Winter is coming, Hagen." I said.

Over the next few days we saw many more formations heading south for the winter. We tired of fishing after an hour and a half and built a small fire on the open gravel and prepared lunch. Back to canned beef stew, the last one from our food store.

It was quiet and very peaceful here by the Ladue. Only the gentlest of breezes stirred tops of trees. The sun shone warmly on our picnic place from its highest point of the day. As we consumed our meal, a few late season mosquitoes wafted around, but the urge seemed to have left them, and they didn't pester us much. After lunch, it wasn't too long before we both dozed, lying on a soft bed of springy grass growing on the yielding peat moss of the river bank.

We woke at two o'clock to the sound of an aircraft engine, It flew low to the north. The plane, a blue and white Piper Super Cub, flew unwavering in a westerly direction. Someone heading toward Tok or Tanacross.

Awake now. Time to try our hand at fishing again. This time I guess I held my mouth correctly because I caught my supper within just a few minutes. Hagen was not far behind with, what I thought, was the same big fish that he caught and released earlier.

"Nice fish, but a slow learner." I mumbled. "Otherwise, you would never have caught it a second time."

Hagen just snorted. From then on it was just for practice and fun. We continued to tease the grayling with our casts. We explored using a variety of flies for the next hour or so.

We quit fishing while the fishing was still good and decided to take our customary medicinal bath in the peat stained Ladue water. Even on the hottest days the Ladue was cold and this late in the season the water was a test for both of us. Well, by now we were supposed to be tough and hardy frontiersmen weren't we? We had goose bumps on top of goose bumps to prove it.

This trial by cold water called for more hot coffee and when we were dry and dressed, we stirred up the fire and boiled water. The sun was still shining so we lounged away a couple more hours before returning to camp. We collected our two fish secured in the cool stream water and headed off.

We traveled the first mile and I noticed a movement at the base of the hill to the right. It was a large bull moose, loping along through the shrubs and trees. I nudged Hagen and pointed, so he throttled back and we watched the moose skirt around the lower flank of the hill, out of sight around the southern tip of the ridge.

"He's right where we made our first airdrop." I said. "I can't believe it's been two years." We really hadn't seen much wildlife so it was a relief to know we didn't drive it all away with our presence in the valley.

With the ATV, it only took twenty minutes to get back to camp; even with the detours, necessary because of thickets and clumps of trees. We entered the camp from the south side.

We had a visitor. The cooking table was upset, the naphtha stove lay on the ground and some of our cooking utensils were scattered around. The canvas front of the cabin was sagging at a crazy angle. It had been pulled from the upper corner support and it looked like the aluminum chest had been pushed from its place.

There was no doubt in our minds of the culprit. Judging by the foot prints all around our camp, the neighborhood black bear was the trickster. It finally got up the courage to come into camp. We probably drove it crazy with the smell of our cooking and it had finally experienced the bear equivalent of a Big Mac Attack. Fortunately, it hadn't broken into the cabin or the aluminum chest so there was no real harm done. Who knows? Maybe we returned with our noisy approach in the nick of time and scared it away.

There were bear paw prints everywhere around the camp and in the mud down by the sluice box. We couldn't tell which way it arrived or departed. We could imagine it lurking in the undergrowth waiting for a second chance. By now we weren't scared of being attacked. We just didn't want it tearing up the camp every time we turned our backs.

We made plenty of noise tidying up and then cooked a delicious smelly meal of boiled potatoes and fish. It must have had the bear drooling something terrible out there in the woods. We almost wished it would show itself so that we could at least take a photo, but it never did.

# Chapter 22
## Last Few Days

Hagen and I spent the next two days working hard to make the very most of our time. Whether it was luck or perseverance I don't know, but we found the most productive zone and extracted the most gold we had done in a two-day period. We also found the largest nugget to date snuggled up against a crevice in the bedrock. It was a beauty, a tear drop shape, and at least a quarter of an ounce in weight.

We were two very tired guys who finally decided it was time to shut down the sluice, carry out the last cleanup for this season, and begin dismantling the camp.

There was a lot to be done, so we had a good night sleep and started about nine the next morning. The sluice was the first to be tackled. We had constructed it from individual modules which unbolted and slipped apart easily. This done, we washed the individual components and set them aside to dry. All of the water hoses were carefully drained, neatly coiled, then tied.

The panning machine was lightweight and we wanted to overhaul the motor and pump during the winter. We'd take it with us.

The cabin would have to come down, so we'd sleep under the stars for the last night. If we didn't take the cabin down, the weight of snow would demolish it. It took us three hours to remove our gear and reduce the cabin to a pile of spars, rolled tarps, and coils of rope.

Some items could be stashed by the rock and inside the enlarged cave. They would survive the winter. Other items would be loaded on the ATV and would be transported out. After lunch we began to move articles to the cache and arrange them neatly as we had the winter before. The smaller, more sensitive items, were shoved to the rear of the cave, while the sturdier items were stacked in front, a barrier to the weather. We were careful not to let the cabin poles or sluice box components rest directly on the ground or they would retain water and go rotten. The items against the cave entrance were covered by a dirty old tarp well staked down around the edges.

A few strategically placed branches and rocks ensured the cache would be overlooked by all but a determined searcher. Once snow arrived the cache would be completely disguised and should be safe.

There was now a large pile of goods stacked by the ATV and we wondered if it would all fit in. We would take our food locker, two footlockers, water pump, panning machine, chain-saw, electric generator, sleeping bags, fishing tackle, weaponry, and many smaller items. There were also six Jerry cans which had held our gasoline supply, now almost exhausted since we topped off the twin tanks of the ATV. Last, but not least, there was a bag containing all of our crushed tin cans, plastic, and other trash. We, unlike some, left nothing of our refuse behind.

We set about loading gear which we would not need for the last night of camping and finished when the light began to fade. We stood back. The camp was no more. One hard working day and it was gone.

We made a small fire and cooked an evening meal from the remaining food supply. We had planned well and would have just a few items for the trail and very little left over.

Dinner completed, we stoked up the fire and lounged

around a while reminiscing about the last few weeks. No patio chairs now. We had mixed feelings about leaving the claim which had been our home for almost four weeks. The excitement of gold mining tended to wear thin after a while and it became almost a chore to dig, sluice, dig, sluice, day after day. There was, on the other hand, the wilderness with its own brand of beauty. We enjoyed days of relaxation now and then. We had the satisfaction of knowing we could survive with only a few civilized comforts. In some ways however, we would be glad to get back to civilization, bank the rewards of our labors and settle down to the normal work routine.

Winter was not far away with downhill and cross country skiing. Yes, there was an interesting variety to our bachelor life in Alaska and we were fortunate enough to experience many facets. There were times when it seemed there were just not enough free days to do all of the things we liked.

At ten, we bedded down by the side of the ATV. Especially, it seemed, for our last night there was a bright three-quarter moon and a frosty nip in the air. We chatted for a while enjoying the warmth of our sleeping bags, and our last night of soothing babble of the little stream which yielded up so much of its wealth. What would happen if the bear returned? No answer and soon we were both snoring.

## Chapter 23

### Back to Civilization

Morning arrived. A light frost rested on the tips of tree branches around the claim and a delicate feathering of ice was on the edges of puddles near the stream. Getting up and dressing was chilly and was accomplished quickly. It surely convinced us we were getting out just in time to avoid the weather change.

In the early light, the claim looked bleak. We made it look lived in for the past four weeks, but now, with everything disassembled and packed away, it was as though there had never been a camp at all. Of course, signs remained of our activity. It was hard to disguise the big excavation in the gravel, the settling pond surrounded by gravel berms, the bridge, our various pathways trampled in the earth. Most of these would, however, be softened by the coming winter.

Sitting around a small cooking fire, we had a breakfast, made for the most part, of leftovers. Then, when it came time to pack away the last of the gear, Hagen recovered the metal box of gold from the secret hideaway. With exaggerated Scrooge-like chuckles, he tucked it safely between the seats of the ATV. It was very rewarding to note how heavy it was and we looked forward to the excitement of carefully sorting and weighing the gold before taking it to a place in Anchorage where they purchased raw gold at spot prices.

Hagen started the engine and drove through the stream to the level ground and left the engine running to warm up while we took a last look around the claim. We wanted to make sure we hadn't left anything lying around.

Satisfied, we mounted the vehicle and drove toward the trail. Hagen paused a moment, stood as high as he could on the ATV and said, "okay Blackie. Take care of it. Its all yours, until next summer."

The ATV was a sturdy machine with a forty-eight horse power, four cylinder power plant; matched with a five-speed gear box. It really had no problem with the trail up to the top of the ridge. We made it easier yet, by choosing a long zigzag route which avoided the steeper slopes. It was still quite a climb, however, and we were always careful not to over tax the engine or transmission. A breakdown at this geographical point would be nothing short of a catastrophe.

Loaded as it was, it took the machine more than an hour to reach the top. As usual, we were extremely careful for the last few yards leading onto the main trail. Hagen drove carefully and avoided maneuvers that would disturb the shale surface or flatten any shrubs. No need advertising our trail entrance. At this place, the surface was firm and didn't readily show tire tracks. The trail to Trib 1 was not visible from the main trail, so if the entrance could remain secret, so could the rest of the trail. Of course, it could be seen from the air but we couldn't help that.

After the long haul up the slope, we gave the ATV a rest at idle and let it cool down. We checked the trail entrance and straightened the few bent over shrubs. This was about as much as we could do to protect our trail from passersby.

Eager to get home now we had started, we didn't dally any longer. We had thirty-five miles to go to the end of the trail, and, barring any abnormal difficulties, it would take us six or seven hours. The ATV was capable of twenty miles per hour but on this terrain we could only average six. Some of the steep sections required extreme care, and coming in fully loaded, we had even used the winch in a couple of places. The outgoing load was not too heavy so we expected to have an easy journey.

Hagen was usually very demanding of machinery and he always expected it to operate at peak. Though the ATV was customized in many ways, it was still a stock machine in terms of the engine, transmission, and suspension. Even Hagen realized he couldn't push this machine too hard because if it broke we were in deep doo-doo. So he treated Herman with the utmost care, used the five gears skillfully and thus rarely needed to apply full power.

We never looked upon speed as the criteria in choosing the vehicle; reliability, load capacity and stability were the main features we sought. Naturally, even at five or six mph, it was far, far quicker than hiking could ever be.

We had no problems over the first fifteen miles which comprised high, dry tundra. When we reached the top of Mt. Son, we stopped for a snack and coffee. Whenever we passed this point, we stopped and enjoyed a great panoramic view of the Ladue Valley and south to the Tetlin. It was a clear day, with a cool breeze, and the view was spectacular. We saw fresh snow dusting upper elevations of the mountains to the west. Termination dust signaled winter was getting closer. As if to emphasize this fact, a couple of flights of geese flew in V formation high overhead.

When we had traveled in almost four weeks ago, the tops of the ridges were lush and green. The patches of white caribou moss, white lichens and splashes of color from a multitude of alpine flowers were great contrasts. Now the ridges had turned a beautiful rustic brown with autumn colors. The bearberry and blueberry bush leaves were brilliant red, yellow, orange, and purple.

From our lofty perch on the mountain, the view to the south was all wilderness. There were no obvious signs of human intervention other than the trail we were following. Way off in the distance to the east, we could make out the lower ridges separating the Trib Valleys. It wouldn't be too long before the whole area we surveyed would be locked up under ice and snow by another long winter.

Scouting around the stunted conifers, we found enough fuel to make a small fire to boil water for coffee. We were soon huddled by the side of the ATV, enjoying the expansive view but sheltered from the keen edge of the breeze. It was good to have a view after being closed in the valley for so long.

Soon it was time to move on. As usual we made sure the fire was out and cleared up our bits and pieces. Over on the far side of the plateau the dozens of rusting forty gallon drums still lay around as evidence of someone else's negligence.

On our trip in with the ATV, we had a very nasty experience. During the scaling of the shale slide, on the western side of the peak, the heavily loaded ATV almost got away from us and we had to blaze an easier way.

The alternate route left the main trail in a northerly direction and then made a wide loop to climb to the flat summit from the north. There was still a steep section but it proved much easier. It also turned out to be an interesting diversion. Part of the north side of the peak featured a picturesque rock cliff about thirty feet in height. The foliage at the foot of the cliff was completely different to that on the top of the mountain, and of course, there was the inevitable bear trail.

It was my turn to drive and I carefully drove down a short steep slope and followed our trail through the shrubbery along the base of the cliff. Our trail gradually dropped to the lower level, and curved around to join the main trail again.

A half dozen black and white spruce hens dodged along the trail ahead of us, as if to say "so long." The ATV was almost upon them before they ran a short distance. They repeated the whole performance. By the time they developed the good sense to take cover at the side of the trail, the brood was scattered a quarter mile along the trail.

The ATV had it easy for the next few miles of level,

straight, trail which traversed a low saddle. Along the way we passed Poplar Hollow where we searched for water two years ago, and had discovered not only water but the strange grove of poplar trees and the animal bones. We revisited the grove on our second hike in, again to get water. We were probably the only ones aware of its existence.

The trail now began its long steady climb toward the next high point. It wasn't named on the map so we called it Squirrel Peak because of the pika squirrels living below the south bluff.

As the trail steepened, I shifted to a lower gear and the ATV settled to a slow but determined pace. Whether the trail was rough or smooth, the ATV kept chugging. There were a few larger washouts to avoid but the rest of the trail was fair. The next couple of miles took quite a long time but eventually the tenacious machine carried us up the last short, steep section and onto the plateau.

I steered the ATV over to the south side and brought it to a halt near the edge of the escarpment overlooking the Ladue Valley. The squirrels were still in residence, but they scurried quickly into their burrows. It wasn't long however before curiosity got the better of them and they were out again, warily keeping a beady eye on us. We stopped several times at this place and must by now have observed several generations of the pika squirrel family.

We stopped just enough time to stretch our legs a bit and to let the ATV engine cool down after the long haul. Riding the machine was, of course, so much easier than hiking, but it brought its own set of problems, and we always completed a trip feeling as if we had been severely bumped and jarred and twisted around.

The trail, from this point, switch-backed up and down for a couple of miles and then climbed gently to the true high point shown on our maps as three thousand eight hundred and six feet. This was a great section to travel because the trail followed a narrow spine with the land falling sharply on both sides. It gave us unrestricted views of the Dennison River Valley to the north and the Ladue and Tetlin area to the south.

More geese to the northwest showed white against the gaunt grey flanks of Mt. Fairplay. They were following the Dennison River Valley but climbing for more altitude. We guessed they headed for the Copper Valley, a major marshaling point, before starting their long migration south.

It was mid-afternoon when we reached the last high point and had our first glimpse of the Taylor Highway about three and half miles away. It would take time for us to pick our way downhill because the trail was so deeply eroded in many places. Thus, we decided to have a late lunch up here where the view was so grand and then tackle the remainder of the trail.

We planned this last meal on the trip by setting aside a couple packages of dehyd beef stew and some crackers. A small fire, some boiled Trib 1 water, and we were soon tucking into our meal accompanied by a mug of sweet tea.

We rested for about an hour, then decided to get going again. The trail ahead was all downhill starting with a difficult shale slide which, unlike the one at Mt. Son, could not easily be bypassed.

To help counterbalance the ATV, Hagen climbed onto the back bumper where he could hang outward as much as possible and shift the center of gravity. I engaged crawler gear, and with the engine almost at idle, started forward to the slope. The idea was to go as slowly as possible and to try to prevent the wheels from locking and sliding on the loose shale.

The ATV tilted down at an alarming angle with Hagen hanging desperately on the back. From the drivers seat, the sloping shale looked truly awesome as it stretched downward for about two hundred feet to end at a jumble of larger rocks. The engine of the ATV, trying to act as a brake, increased speed but I held the rpm down by careful application of the brakes. The wheels locked and skidded when a patch of shale broke loose, but I managed to keep the nose pointing downhill and get the wheels rolling again so they found a fresh purchase.

It made for a hypertensive ride, but, we made it safely to the bottom of the slope. That was the important thing. I stopped briefly to let Hagen retake his front seat and then continued down the more gentle grade. It was easy going, but we kept a good look out for eroded areas which are not so easy to spot when traveling downhill. Some more severely eroded places could easily have upset the ATV. Except a few places, where Hagen clung to the back or the side, we skirted around most of the serious washouts.

The view over the tree tops was colorful, with the fall leaves sharp and vivid. The deciduous trees, with their yellow, orange, and red blended and contrasted sharply with the dark green of the Sitka and black spruce. The scene over the valley ahead was brilliant and made us squint. In this area, fall colors lasted for a brief time so we were lucky to catch them at their very best.

After forty-five minutes of ear popping descent, the trail leveled out. We reached the swamp, a barrier, three hundred yards or so from the end of the trail. It wasn't so much of an obstacle to the ATV, but it muddied up an otherwise clean vehicle. Without hesitation, I drove slowly through to the dry ground. The water wasn't too deep. There had been very little rain in past weeks.

The Chevy and trailer were exactly as we had left them hidden back among the trees. Taylor Highway was so close, we worried in case the vehicle was discovered. It wouldn't take too much for someone, who knew we were working out on the claim, to figure that we might have a vehicle stashed somewhere at trail's end. Thus we expended a lot of effort in concealing the Chevy every time we made these trips. It would have been terrible to return and find the vehicle stripped of parts or even missing altogether.

We left the ATV on the main trail and set about removing the camouflage from the Blazer. The battery was a bit low on energy but the engine, now with ninety-five thousand miles on it, groaned to life after a few cranks. We gave it a few minutes to warm up, and folded the tarps and placed them in the back. It took a bit of maneuvering to get the vehicle and trailer out to the trail. As usual, we tried to do it with minimum disturbance to the trees and bushes.

Watching out for traffic, we carefully guided each vehicle in turn over the berm, onto the highway and up to the gravel pit. Only then did we load the ATV onto the trailer. It was preferable to wrestling the loaded trailer over the berm.

It took us a few minutes to transfer important items, gold and rifles, from the ATV to the Blazer. A few minutes more, the ATV was secure on the trailer with specially positioned chains.

The year before, when we exited the trail, we had heard a rattling sound. An older couple on heavily loaded, bicycles came along the gravel road. Bicycles were the last thing we expected to see on the Taylor Highway. We started a conversation with the riders, an elderly Swiss man and his wife. They had cycled from San Francisco, taking several months, roughing it and camping by the roadside when there was nothing more convenient. They were now on the final, easy, leg to Anchorage for a flight back home. We had to admire their spirit for adventure and were pleased to be able to give them a rundown on road conditions to Anchorage. Smooth going for them after the two hundred miles of Taylor Highway gravel and goodness knows how many miles on the Al-Can.

No such excitement this year. While we were loading the ATV, a couple of motor homes rattled down the highway, not slowing, left us choking in dust. Welcome again to civilization. One last thing, with a shovel from the Chevy, we restored the berm at the trail entrance. There was no point in advertising its presence.

Just in case we ran into foul weather on the way home, we lashed a tarp securely over the ATV. With that we were ready to move out. It was five o'clock in the evening of an easy day. A good meal at the restaurant in Tok, and we would be all set to take turns driving to Wasilla.

I drove fairly slow for the few miles to Tetlin Junction and then turned west on the asphalt of the Al-Can.

Twenty minutes later we parked along with a variety of motor homes and campers in front of the Grizzly Restaurant. We chose a table by the windows and order hamburgers, french fries, and coffee. While we were waiting, we took turns using the facilities to freshen up.

With improved facilities at camp, like hot water and a shower when we needed it, we remained well groomed and tidy so we didn't seem at all out of place among other patrons. It was the tail end of tourist-season and most of the people were on the road.

We talked as we ate, and typically, got around to what we'd do differently next year. We hadn't even gotten home from this trip and we were already planning for the next mining season.

Our meal finished, we had the waitress fill our flask with hot coffee, paid the tab, and stepped outside. We cast an eye around the Chevy and the trailer, decided that everything looked secure and climbed aboard.

I eased the rig out of the parking lot, turned west, and accelerated to a comfortable fifty-five. Watch out for those frost heaves in the road surface. There were almost three hundred miles to go.

Ahead of us the sun was already angling toward the high peaks of the Mentasta Range. A large gaggle of geese, black against the tinted sky, slid down on motionless crescent wings toward some hidden lake or river south of the highway making an overnight stop on their way south. They'd return next year. Winter would have a mean, icy grip on this country and it would be no fit place for placer miners or geese. After spring-breakup, life would move back in for another full cycle.

Such was the lure of wild Alaska. Once it was in your blood, you were stuck. The presence of gold in Trib 1 was only a part of the equation, an excuse for harmony with the wilderness and to partake of its offerings. Even if you left, there was always some hidden force beckoning you back. We were not much different to those geese. Just as surely, we would be drawn back again when the season was right.